AF794808

ASCE Manuals and Reports on Engineering Practice No. 115

Pipe Ramming

Second Edition

Prepared by the
Pipe Ramming Task Force of the
Trenchless Installation of Pipelines (TIPS) Committee
of the American Society of Civil Engineers

ASCE AMERICAN SOCIETY OF CIVIL ENGINEERS

UESI ASCE

Published by the American Society of Civil Engineers

Library of Congress Cataloging-in-Publication Data

Names: American Society of Civil Engineers, author.
Title: Pipe ramming / Prepared by The Pipe Ramming Task Force of the Trenchless Installation of Pipelines (TIPS) Committee of the American Society of Civil Engineers.
Description: Second edition. | Reston, Virginia : American Society of Civil Engineers, [2020] | Series: ASCE Manuals and Reports on Engineering Practice ; No. 115 | Includes bibliographical references and index. | Summary: "MOP 115, Second Edition, describes current pipe ramming practices in design and construction of pipelines under roads,
railroads, streets, and other constructed and natural structures and obstacles and includes information on culvert replacements, pipe crushing, slickbores, pipe extraction, as well as assists with horizontal directional drilling projects"-- Provided by publisher.
Identifiers: LCCN 2020012202 | ISBN 9780784415603 (paperback) | ISBN 9780784483022 (adobe pdf)
Subjects: LCSH: Pipelines--Design and construction. | Trenchless construction.
Classification: LCC TA660.P55 P54 2020 | DDC 621.8/672--dc23
LC record available at https://lccn.loc.gov/2020012202

Published by American Society of Civil Engineers
1801 Alexander Bell Drive
Reston, Virginia 20191-4382
www.asce.org/bookstore | ascelibrary.org

Any statements expressed in these materials are those of the individual authors and do not necessarily represent the views of ASCE, which takes no responsibility for any statement made herein. No reference made in this publication to any specific method, product, process, or service constitutes or implies an endorsement, recommendation, or warranty thereof by ASCE. The materials are for general information only and do not represent a standard of ASCE, nor are they intended as a reference in purchase specifications, contracts, regulations, statutes, or any other legal document. ASCE makes no representation or warranty of any kind, whether express or implied, concerning the accuracy, completeness, suitability, or utility of any information, apparatus, product, or process discussed in this publication, and assumes no liability therefor. The information contained in these materials should not be used without first securing competent advice with respect to its suitability for any general or specific application. Anyone utilizing such information assumes all liability arising from such use, including but not limited to infringement of any patent or patents.

ASCE and American Society of Civil Engineers—Registered in US Patent and Trademark Office.

Photocopies and permissions. Permission to photocopy or reproduce material from ASCE publications can be requested by sending an email to permissions@asce.org or by locating a title in the ASCE Library (https://ascelibrary.org) and using the "Permissions" link.

Errata: Errata, if any, can be found at https://doi.org/10.1061/9780784415603.

Copyright © 2020 by the American Society of Civil Engineers. All Rights Reserved.

ISBN 978-0-7844-1560-3 (print)
ISBN 978-0-7844-8302-2 (PDF)

Manufactured in the United States of America.

26 25 24 23 22 21 20 1 2 3 4 5

Cover photo credits:
Front left: Courtesy of Hammerhead Trenchless
Front top right: Courtesy of TT Technologies
Front bottom right: Courtesy of TT Technologies
Back top: Courtesy of Claude H. Nix Construction
Back bottom: Courtesy of Hammerhead Trenchless

MANUALS AND REPORTS ON ENGINEERING PRACTICE

(As developed by the ASCE Technical Procedures Committee, July 1930, and revised March 1935, February 1962, and April 1982)

A manual or report in this series consists of an orderly presentation of facts on a particular subject, supplemented by an analysis of limitations and applications of these facts. It contains information useful to the average engineer in his or her everyday work, rather than findings that may be useful only occasionally or rarely. It is not in any sense a "standard," however, nor is it so elementary or so conclusive as to provide a "rule of thumb" for nonengineers.

Furthermore, material in this series, in distinction from a paper (which expresses only one person's observations or opinions), is the work of a committee or group selected to assemble and express information on a specific topic. As often as practicable the committee is under the direction of one or more of the Technical Divisions and Councils, and the product evolved has been subjected to review by the Executive Committee of the Division or Council. As a step in the process of this review, proposed manuscripts are often brought before the members of the Technical Divisions and Councils for comment, which may serve as the basis for improvement. When published, each manual shows the names of the committees by which it was compiled and indicates clearly the several processes through which it has passed in review, so that its merit may be definitively understood.

In February 1962 (and revised in April 1982), the Board of Direction voted to establish a series titled "Manuals and Reports on Engineering Practice" to include the manuals published and authorized to date, future Manuals of Professional Practice, and Reports on Engineering Practice. All such manual or report material of the Society would have been refereed in a manner approved by the Board Committee on Publications and would be bound, with applicable discussion, in books similar to past manuals. Numbering would be consecutive and would be a continuation of present manual numbers. In some cases of joint committee reports, bypassing of journal publications may be authorized.

A list of available Manuals of Practice can be found at http://www.asce.org/bookstore.

CONTENTS

PREFACE IX
ACKNOWLEDGMENTS XI

1. INTRODUCTION 1
References 5

2. PIPE RAMMING PROCESS 7
2.1 Introduction and Background 7
2.1.1 Introduction 7
2.1.2 Background 9
2.2 Method Description 11
2.3 Equipment Selection and Setup 13
2.3.1 Guide Rails and Setting Grade 13
2.3.2 Minimum Space Requirements 14
2.3.3 Depth 16
2.3.4 Hammer Selection 19
2.3.5 Air Compressor 20
2.3.6 Cutting Shoe 21
2.3.7 Casing Selection and Wall Thickness 24
2.3.8 Hammer Attachment 26
2.3.9 Efficiency of Casing Penetration or Advancement... 29
2.3.10 Lubrication 32
2.3.11 Joining Casing Pieces 33
2.3.12 Spoil Removal 36
2.3.13 Carrier Pipe Installation 36
2.4 Pilot Tube Guidance 41
2.4.1 Pilot Tube Installation 41
2.4.2 Step-Up Casings and Installation Sequence 42
2.5 Typical Applications 42
2.5.1 Road and Railroad Crossings 42
2.5.2 Culverts 44

2.5.3 Pipe Crushing 46
2.5.4 Slick Boring 47
2.5.5 Pipe Extraction 48
2.5.6 Combining Pipe Ramming and Horizontal Directional Drilling 49
References 54

3. PLANNING 55
3.1 Initial Criteria 55
3.1.1 Casing Diameter and Grade 56
3.1.2 Depth 57
3.1.3 Routing 58
3.1.4 Portals, Shafts, and Working Space 59
3.1.5 Sustainability 62
3.1.6 Regulations 63
3.2 Site Conditions 63
3.2.1 Land Use 63
3.2.2 Easements and Rights-of-Way 64
3.2.3 Underground Utilities and Other Facilities 65
3.2.4 Buried Objects 66
3.2.5 Contaminated Ground 66
3.2.6 Special Considerations 67
3.3 Preliminary Geotechnical Considerations 70
3.4 Project Layout 71
3.4.1 Layout for Efficient Pipe Ram 71
3.4.2 Alignment and Distance Considerations 72
3.4.3 Effects on Adjacent Structures 72
3.5 Cost Considerations 73
3.5.1 Direct Cost of the Pipe Ramming 73
3.5.2 Indirect Social Effect Costs 77
3.5.3 Environmental Costs 80
3.5.4 Risk Analysis 81
3.5.5 Contingency Costs 82
References 83

4. DETAILED SITE INVESTIGATIONS 85
4.1 General 85
4.2 Geotechnical Assessment 85
4.2.1 Site Investigation 86
4.2.2 Characterization of Subsurface Materials 90
4.2.3 Geotechnical Reports 95
4.2.4 Applicability of Pipe Ramming Based on Subsurface Conditions and Classification 96
4.3 Utility Surveys 96
4.4 Traffic Flow and Access for Vehicles and Pedestrians 98

4.5 Environmental Conditions ... 99
4.6 Flood Zones ... 100
4.7 Seismic Considerations ... 101
References ... 102

5. DESIGN ... 105
5.1 General ... 105
5.2 Planning and Permitting ... 106
5.2.1 The Process ... 106
5.2.2 Cost and Schedule ... 107
5.2.3 Hazard Identification ... 108
5.2.4 High Risk Hazards ... 111
5.2.5 Permitting and Design for Special Crossings ... 111
5.3 Exploration ... 112
5.3.1 Planning an Exploration Program ... 112
5.3.2 Site Layout ... 112
5.3.3 Geotechnical Considerations ... 114
5.3.4 Groundwater ... 115
5.3.5 Ground Settlement or Heave ... 115
5.3.6 Deviations in Line and Grade ... 116
5.3.7 Appropriate Casing Size ... 118
5.3.8 Ram Length Limitations ... 118
5.4 Design Considerations ... 119
5.4.1 Basis of Design ... 119
5.4.2 Selection of Line, Grade, and Tolerances ... 119
5.4.3 Casing Material Selection and Dimensions ... 120
5.4.4 Ground Response (Settlement, Heave, and Vibration) ... 121
5.4.5 Joint Configuration ... 123
5.4.6 Driving Resistance of the Casing ... 124
5.5 Documentation ... 125
References ... 127

6. CONSTRUCTION ... 129
6.1 General ... 129
6.2 Bidding ... 129
6.3 Submittals ... 130
6.3.1 Preconstruction Submittals ... 131
6.3.2 Construction Submittals ... 134
6.3.3 Postconstruction Submittals ... 135
6.4 Measurement and Payment ... 135
6.5 Jobsite Layout and Equipment Setup ... 138
6.6 Survey ... 139
6.7 Launching and Receiving Shafts ... 140
6.8 Steel Casing, Leading Edge Treatments, and Welding ... 142

6.9 Safety Issues ... 143
6.10 Differing Site Conditions ... 143
6.11 Traffic Control, Fencing, and Barricading ... 144
6.12 Quality Control ... 144
6.13 Ramming Forces and Lubricants ... 145
6.14 Spoil Removal and Disposal ... 146
6.15 Inspection and Monitoring ... 147
6.16 Reports and Records ... 148
6.17 As-Built Drawings and Documentation ... 149
References ... 149

7. SUMMARY OF RECENT CASE HISTORIES ... 151
7.1 General ... 151

GLOSSARY ... 179
INDEX ... 193

PREFACE

Manual of Practice No. 115, Second Edition, was prepared by the Pipe Ramming Task Force of the ASCE Committee on Trenchless Installation of Pipelines (TIPS) as part of the Utility Engineering & Surveying Institute (UESI). The manual describes current pipe ramming practices used by engineers and construction professionals in designing and constructing pipelines under roads, railroads, streets, and other constructed and natural structures and obstacles.

This manual has been created by a group of engineers, owners, suppliers, manufacturers, and contractors fully knowledgeable of the method and its use. This manual considers many of the advances that have occurred over the years with pipe ramming. Many of the sections provide a summary of the state of the industry as of 2019. The task force acknowledges that the technology continues to change and that changes in construction continue to develop.

Sections have been written assuming the reader may be new to the various construction methods included in this manual. No document including this one can encompass all the issues on a particular pipe ramming project. Improvements in best practices and technology continue to evolve so quickly that consideration of this manual on any project must take into account not only the specific characteristics of the particular project but also further improvements in best practices and technology.

The engineer of a pipeline is encouraged to consider all trenchless methods before concluding that pipe ramming is the most suitable construction method available. Manuals and Reports on Engineering Practice (known as MOPs) have been written by ASCE for different construction methods. A list of useful references specific to the various trenchless methods is provided at the end of Chapter 1.

If the engineer responsible for the pipeline project does not have a strong background in trenchless design, an engineering firm that specializes in trenchless designs should be consulted to provide a peer review early in the planning/design process to ensure good design choices are being made.

ACKNOWLEDGMENTS

ASCE and the Utility Engineering & Surveying Institute (UESI) acknowledge the work of the Pipe Ramming Task Force. The group comprises individuals from many backgrounds, including consulting engineering, the construction industry, equipment and pipe manufacturing, education, government, design, and private practice.

Principal authors of this manual of practice (second edition) are

Glenn M. Boyce, Ph.D., P.E., P.Eng, *Task Committee Chair*
D. Craig Camp
Brian Dorwart, P.E.
Alan Goodman
Michelle L. Macauley, P.E.
Kevin P. Nagle
Kim Staheli, Ph.D., P.E.

Other individuals who served and contributed on the manual of practice are

George Davis, P.E.
Brent Keil
Armin Stuedlien, Ph.D., P.E.

Thank you to the Blue Ribbon Review Committee who reviewed the finished document and provided their comments:

Mohammad Najafi, Ph.D., P.E., *Blue Ribbon Committee Chair*
Harry Dickinson
Dennis J. Doherty, P.E.
Rick Melvin

Thank you to Julie McCullough for providing technical editing and Steve Morrison and Jon Clugston for graphic support.

The task force would like to acknowledge the financial support provided by the following companies and associations:

Brierley Associates
GHD
Hammerhead Trenchless
Macauley Trenchless
McMillen Jacobs Associates
TT Technologies

ASCE Committee on Trenchless Installation of Pipelines (TIPS)

Chair: Jeffery J. Boschert, P.E.
ExCom Liaison: Tennyson Muindi, P.E.

The authors would also like to recognize and thank the contributors of the first edition (2008) of MOP 115:

Mohammad Najafi, *Chair*
Craig Camp, *Secretary*
Larry Slavin, *Vice Chair*
Jeff Bikshorn
Fred Burlbaw
Robert Carpenter
George Davis
Dennis Doherty
Ahmad Habibian
John Hair
Baosong Ma
Terry McArthur
Dave Mittler
Terry Moy
Kevin P. Nagle
Collins Orton
Mike Schwager
Mike Sechelski
Hanson Turnbull

CHAPTER 1
INTRODUCTION

Communities around the world routinely install underground utilities of various types, sizes, and purposes. These utilities form the infrastructure that provides our homes, office buildings, schools, and other structures with

- Stormwater drainage,
- Waste removal via sanitary sewers,
- Potable water,
- Recycled water,
- Irrigation,
- Oil and gas (transmission and distribution),
- Electrical power (transmission and distribution),
- Communication, and
- Cable.

These pipelines are installed anywhere and by a variety of construction methods, with the most common being open trench excavation. Municipal infrastructure is typically installed within the rights-of-way of our roads, streets, and highways. Cross-country utilities, such as oil and gas pipelines and power transmission lines, are typically installed in their own rights-of-way. As new pipelines are placed below the surface and some utilities are abandoned, the density of the underground infrastructure increases. As areas are built up and the amount of traffic increases on the road surfaces, access to the subsurface correspondingly decreases. Therefore, putting new utilities and pipelines within an existing right-of-way has become more difficult, and fewer utilities can be installed by open trench excavation. The construction industry has adapted to meet the challenges of installing pipelines in congested, urban environments around the world and now offers increasingly sophisticated solutions for both rehabilitating existing and installing new infrastructure. Because these new construction solutions/methods do not rely on digging a trench, they are known as trenchless methods.

Some of the more widely used trenchless methods for new utility installations are (listed alphabetically)

- Guided boring (pilot tube or front steer),
- Horizontal auger boring (HAB),
- Horizontal directional drilling (HDD),
- Microtunneling,
- Pipe bursting,
- Pipe ramming
- Shielded pipe jacking, and
- Utility tunneling.

Trenchless methods are often attractive to project owners and are sometimes required by regulatory agencies and property owners because they reduce impacts to the surface, existing structures, and infrastructure. Above-ground effects are reduced because excavation is limited to shafts, pits, or portals, rather than a continuous trench. Underground effects are reduced because the pipe can be installed without excavating the otherwise conflicting underground utilities or surface structures. In general, the benefits of trenchless construction methods can be summarized as follows:

- Provide improved public safety;
- Lower social and economic effects;
- Minimize environmental effects, including carbon footprint;
- Reduce impact of construction activities on traffic;
- Allow for deeper installations without significant increases in the cost;
- Mitigate impacts to existing surface and subsurface infrastructure;
- Address potentially challenging subsurface conditions such as soft or loose soil, rock, or the groundwater table;
- Allow more freedom in alignment selection within or outside of established rights-of-way; and
- Reduce project duration.

However, trenchless installations are much more sensitive to subsurface conditions than open trench excavations. The active excavation face has limited access. Construction is completed from within the tunnel or pipe as a confined space. Cutting tools must be selected in advance and predictions made on how the ground will behave during excavation and removal. Also, it is usually more difficult, costly, and time-consuming to adapt trenchless methods to unexpected ground conditions than it is to adapt open trench excavations. This is primarily because when trenchless methods are used, the material being excavated is not always visible or directly accessible at the face. Therefore, before choosing a specific trenchless method, it is critical to understand the subsurface ground conditions and the interaction of cutting tools with the ground. Because the ground is not an engineered material with known properties and extents, it becomes imperative to conduct a thorough geotechnical and subsurface investigation to determine, among other things, geology, ground type and anticipated behavior, groundwater level, and presence of buried objects placed by natural means or as the result of past construction. The anticipated conditions affect the selection of the trenchless method, design layout, cutting tools, and contract provisions.

The basic design process for a trenchless project should include identifying and understanding

- Project requirements,
- Owner design requirements,
- Third-party requirements,
- Ground conditions, and
- Risks and risk tolerance.

This information is then used in the design phase to

- Select a suitable trenchless method,
- Provide a design that manages the risks,
- Mitigate the controllable risks and assign responsibility for the other risks,
- Identify potential claims and minimize them, and
- Prepare contract clauses that help allocate risks and manage disputes.

The decision to use a trenchless method, and the selection of a particular method from those available, should be made methodically and be systematically based on project requirements. Trenchless methods, equipment, and guidance systems improve and change over time, and contractors continue to develop innovative new techniques. Therefore, it is important for design engineers to keep abreast of available trenchless methods and tooling to systematically evaluate options. Regardless of the trenchless method selected, adherence to the following guidelines can reduce project cost, schedule, and risk:

- Use uniform outside diameters (ODs) to the extent possible,
- Use similar drive lengths to the extent possible,
- Keep drive lengths within practical limits,
- Minimize the number of shafts,
- Colocate shafts at maintenance access points,

- Consider frequency and size of connections, and
- Evaluate alternative alignments.

As noted, there are different trenchless methods available to engineers and contractors to excavate and support the ground during installation. These methods address various pipe materials, sizes, and lengths, and offer different capabilities as follows:

- Pipe size and pipe material,
- Need for a full or temporary casing,
- Drive length,
- Installation above or below the groundwater table,
- Installation at a recise line and grade for gravity flow applications, and
- Installation of the pipeline directly (one-pass process) or with a casing (two-pass process).

Table 1-1 summarizes the typical pipe sizes, installation lengths, ability to install pipelines above or below the groundwater table, and ability to install pipelines at a precise line and grade for the more common trenchless methods. The pipe sizes and drive lengths reflect advances that have occurred as the industry has matured. However, the emphasis should always be placed on managing risk first, not matching or exceeding lengths accomplished in the past. Actual and perceived risks directly increase an owner's cost.

For information on the other trenchless methods provided in Table 1-1, the reader may refer to the following:

- Horizontal auger boring: ASCE MOP 106 (2017a);
- Horizontal directional drilling: ASCE MOP 108 (2014);
- HDD good practices guidelines: NASTT (2017);
- Microtunneling: ASCE 36-15 (2015);
- Pilot tube and other guided boring: ASCE MOP 133 (2017b);
- Shielded pipe jacking: Thomson (1994); and
- Utility tunneling: Proctor and White (1977).

This manual of practice (MOP) focuses specifically on pipe ramming. The chapters in this manual describe the following:

- Chapter 2: Pipe Ramming Process,
- Chapter 3: Planning,
- Chapter 4: Detailed Site Investigations,
- Chapter 5: Design,
- Chapter 6: Construction,
- Chapter 7: Summary of Recent Case Histories,
- Glossary, and
- Index.

Table 1-1. Typical Sizes, Lengths, and Capabilities of Various Trenchless Methods.

Method	Size, OD (in.)	Length (ft)	Groundwater	Grade control
Pipe ramming[a]	8–96	250	Both[b]	No[c]
Horizontal auger boring[a]	12–72	300	Above[b]	No[c]
Guided boring[a]	4–48	300	Both[b]	Yes
Microtunneling[d]	30–96	1,500	Both	Yes
Shielded pipe jacking[d]	48–120	1,500	Above[b]	Yes
Horizontal directional drilling	4–42	6,500	Both	No
Utility tunneling[d]	48 or larger	Any length	Above[b]	Yes

Note: These are typical parameters and do not represent absolute minimum and maximum, nor do they consider additional ground improvement methods or other risk mitigation methods. These values are subject to change as technology evolves.

[a]Length decreases as diameter increases.

[b]Below the groundwater may be possible, but it is limited by ground type or depth below groundwater.

[c]Limited grade control may be achievable through steering heads or hybrid methods, as discussed herein.

[d]Length increases as diameter increases.

REFERENCES

ASCE. 2007. *Pipe bursting projects*, MOP 112. Reston, VA: ASCE.

ASCE. 2014. *Pipeline design for installation by horizontal directional drilling*, MOP 108, 2nd ed. Reston, VA: ASCE.

ASCE. 2015. *Standard design and construction guidelines for microtunneling*, ASCE/CI 36-15, April. Reston, VA: ASCE.

ASCE. 2017a. *Horizontal auger boring projects*, MOP 106, 2nd ed. Reston, VA: ASCE.

ASCE. 2017b. *Pilot tube and other guided boring methods*, MOP 133. Reston, VA: ASCE.

NASTT (North American Society for Trenchless Technology). 2017. *Horizontal directional drilling (HDD) good practices guidelines*, 4th ed. Cleveland: NASTT.

Proctor, R. V., and T. L. White. 1977. *Earth tunneling with steel supports.* Youngstown, OH: Commercial Shearing, Inc.

Thomson, J. 1994. *Pipejacking and microtunneling*. New York: Chapman & Hall.

CHAPTER 2
PIPE RAMMING PROCESS

2.1 INTRODUCTION AND BACKGROUND

This manual of practice addresses pneumatic and hydraulic-powered pipe ramming, which is widely used for installation of steel casings and cross-country steel pipelines primarily at crossing locations. The manual does not address other types of trenchless technology methods, which are also used for the installation of casings and pipelines, such as pilot tube method, horizontal auger boring, HDD, microtunnneling, pipe bursting, shielded pipe jacking, and utility tunneling (see Chapter 1 for related documents).

2.1.1 Introduction

Pipe ramming can be a cost-effective technique for inserting a steel casing from a launch shaft, portal, or trench by means of driving the casing through ground using a pneumatic or hydraulic percussion hammer, which is attached to the end of the steel casing or steel pipeline.

Pipe ramming is a dynamic process of hammering the casing into the ground. Discussions in this chapter use terms to describe this dynamic process that have specific meanings for clarity. Terms are defined in the Glossary.

Dynamic installation of casings and pipes by the pipe ramming process is analyzed using the actual energy of the hammer system and the efficiency of transferring the rated energy to the casing being driven. Insufficient data exist for the determination of actual hammer energy; therefore, most analyses to date have used the rated energy of the hammer.

A parameter measured during the pipe ramming process is the time it takes to install a casing into the ground, which is called the *penetration rate*. When energy is applied, it must overcome the resisting forces or the casing will not advance. System efficiency is a measure of the amount of available energy that is actually transmitted through the assembly to the leading edge and is available for advancing the system. The available drive energy per blow must be enough to overcome the friction and inertial resistance to move the casing forward. *Work* as a form of energy is equal to force times the change of displacement. For pipe ramming, work may then be assessed either as a time interval required to penetrate a given distance or as a length over a given time interval. Penetration rate and work are the basic elements of estimating cost for pipe ramming projects. Energy is used to select the hammer size, and work is used to estimate how long the actual driving will take for the full installation. Work is based on the penetration rate, which is a function of the drive system efficiency and the interaction of the drive system with the ground being penetrated. Lower drive energy hammers applying a given energy at a higher frequency (blows per minute) can achieve a similar amount of work as a higher drive energy hammer applied at a lower frequency (blows per minute). Basically, it is like driving a pile into the ground.

Pneumatic hammers typically strike at frequencies of 150 to 500 blows per minute, whereas hydraulic hammers strike at lower frequencies, typically in the range of 30 to 60 blows per minute. Ground shear failure is required to advance a casing. Different ground types and densities react differently to different frequencies; however, few field data are available to determine optimal drive frequencies for specific ground types. This feature of hammer/ground dynamics has been extensively studied in both dynamic pile and foundation designs. Results indicate that frequency changes may be used to change the penetration rate and may even be applied to cause the driven casing grade to rise or drop during advancement.

Steel casing is used in essentially all applications because the steel can be designed with sufficient strength. There have been situations in which a hammer has been used to help with or extract high-density polyethylene (HDPE) pipe in an HDD assist, but the transfer of energy is significantly less through HDPE pipe and works on a limited basis. Steel can have a relatively thin section to handle the ramming forces without buckling and can be stiff enough to transfer the maximum amount of drive energy to the cutting shoe. The key design question that then remains is whether the amount of energy at the cutting shoe is enough to penetrate the ground. Steel casings have been installed in lengths exceeding 400 ft (122 m) for diameters up to 60 in. (1,525 mm), and for relatively short distances [e.g., 250 ft (76 m)] with diameters exceeding 120 in. (3,050 mm).

Pipe ramming is most appropriate for installation of large-diameter casings over a short distance and for shallow-depth installations. It is suitable for all ground conditions except those for which the entire casing perimeter is engaged with rock or other high-strength materials (i.e., very dense clay). Pipe ramming installations can be successful where some other trenchless methods can lead to unacceptable surface settlement. Pipe ramming is most commonly used in applications with a horizontal orientation, but it can also be used for inclined and vertical applications, such as installing conductor and washover casings for HDD projects and driving pipe piles and micropiles.

The primary horizontal applications of this method are the placement of casings beneath roadways or railroads for cross-country pipelines or as culverts to pass streams and creeks. The casings may contain water or wastewater pipelines or serve as a utility gallery for electric, telecommunications, or fiber optic cables. The installed casing can also be used directly as a storm culvert or product pipeline in which interior coatings are applied after spoils are removed. Pipe ramming has also been used to install a larger pipe over an existing pipeline, also known as *pipe swallowing* or *pipe crushing*. Pipe ramming for pipe extraction involves pushing one steel pipe section out with a new steel pipe of the same size. Pipe ramming is an economical alternative to open trenching, because it can reduce pavement damage, traffic disruptions, amount of excavated spoils, and social costs associated with pipeline installations. A major social effect from the pipe ramming process is the noise generated by pneumatic driving equipment and associated compressors. This noise issue will be discussed later in this manual.

This Manual of Practice has been prepared to assist engineers, contractors, and owners who are using the pipe ramming method to design and execute pipe installation projects effectively, safely, and in conformance with project requirements and site conditions. The objective of this manual is to give a clear understanding of the method's capabilities and limitations, outline important design and construction considerations, and identify potential problems and prevention measures, thereby instilling confidence in the science and art associated with this method. The guidelines provided are based on case histories, workshops, project reviews, technical papers, and other information contributed by industry experts.

2.1.2 Background

Early pipe ramming project design was based on the experience of the contractors and hammer manufacturers. As pipe ramming projects have become more common, general design guidelines for casing diameter and wall thickness, maximum drive length, shoe design, and minimum hammer size were developed from experience. Engineers have been assessing the data and developing relationships between known hammer properties and ground properties to better predict hammer size for various lengths and diameters in different ground types. Both engineers and pipe manufacturers continue to publish information and case histories correlating casing diameter and wall thickness to drive length and appropriate hammer size for various casing diameters, for example, TT Technologies (https://www.tttechnologies.com/), Hammerhead Trenchless (https://www.hammerheadtrenchless.com/), and IHC Hydrohammer (https://www.ihciqip.com/en/products/piling-equipment/hydrohammer). Chapter 7 provides a listing of case histories that may be useful in the assessment of pipe ramming means and methods for a specific project.

When pipe ramming was introduced, the steel casings were primarily installed as culverts for crossing critical structures such as railroads or highways. In general, pipe ramming was used for the installation of casings of small diameters from 6 to 12 in. (152 to 305 mm), with drive lengths of less than 65 ft (20 m), and only above the groundwater. However, the length and diameter of rammed crossings continually increased as the pipe hammer capabilities increased. Crossings are now successfully installed at diameters as large as 144 in. (3,360 mm) with lengths of more than 250 ft (76 m), and at smaller diameters with lengths of more than 500 ft (152 m); however, these are the extreme cases. (See Chapter 7 for a table of recent case histories.) Innovations in pipe ramming have allowed successful installation of pipe rams beneath the groundwater with proper use of temporary plugs at the face of the ram to resist flowing ground conditions.

Pipe ramming design is not well established, and design elements remain under development. Early designs consisted of using a bigger hammer if a casing became stuck or using thicker casing to prevent buckling. As contractors and hammer manufacturers continued to push the limits of pipe ramming, new information and observations have resulted in empirically based design suggestions being developed by the manufacturers that were closely restricted within the manufacturers' ranks. Little if any geologic or geotechnical information was obtained. Manufacturers typically would provide suggested guidance for casing wall thickness and recommend the hammer size. Formal designs were not done, simply because they did not add value to the simple road crossing projects. However, as the economics and speed of the technique increased, more complex projects were attempted. Various types of collateral issues started to become evident, such as heaving of the ground surface, damage to adjacent utilities, and misalignment of the casing. These issues may be mitigated by the design and understanding of the ground reaction to the pipe ramming process. Although it is the associated risk and risk ownership that have increased the cost of pipe ramming projects, some design methods now provide significant added value and should be considered.

The critical design parameters are driven by the specific application needed on a project. Most often, steel pipes installed with ramming are used as casings that house smaller diameter carrier pipes that are subject to pressure or gravity flow. Pipe ramming is also routinely used to install conductor and washover casings for HDD projects. However, depending on numerous factors, the designers may choose to use the casing as the carrier pipe, as in the case of a culvert or installation of a pipeline in a slick bore application (Figure 2-1).

2.2 METHOD DESCRIPTION

The two major categories of pneumatic or hydraulic pipe ramming are *closed-face* and *open-face*. Closed-face pipe ramming (also called *piercing, impact moling*, or *the compaction method*) was the original method of dynamic installation of pipes and utilities. The dynamic process was adapted from pile driving with equipment capabilities to provide a horizontal blow with enough energy to drive a small-diameter steel pipe. The closed-face process includes welding a cone-shaped head to the leading end of the first segment of casing. Note that the open-face process can become a closed-face process if the ground becomes plugged within the open-faced casing,

Figure 2-1. Slick bore casing being positioned in a trench for pipe ramming into the ground.
Source: Courtesy of Hammerhead Trenchless.

preventing the ground from entering the casing. In the closed-face process, the leading cone penetrates and displaces the surrounding ground as the casing is rammed forward by a hammer placed at the rear of the casing. The ground-casing installation interaction with this technique is similar to what occurs during soil-compaction methods. The ground is compressed and displaced as the system advances and creates a cavity for the trailing pipe. This technique can be used for casing up to 7 in. (180 mm) in diameter. This ground displacement process may cause ground heave or may apply load to adjacent utilities. A zone of influence is typically considered to identify the geometry for potential high-risk damage. Experience has shown that damage to crossing utilities and ground heave are a function of depth and density of the ground. Experience gained from the use of the pipe bursting method may provide some guidance in the determination of a potential damage zone. Pipe bursting typically recommends having 1 ft (0.3 m) of cover for each inch (25.4 mm) of diameter of the constructed cavity and a lateral clearance of a minimum of two pipe diameters from other utilities. Heave can occur at shallow depths with the closed-face pipe ramming. Closed-face casing has been used on nonengineered projects like irrigation lines, house connections, and sewer laterals. Use of a closed-face pipe ram eliminates the need for spoils removal.

An open-face pipe ramming process allows the front end of the lead casing to act as a "cookie cutter" or cutting ring, thus allowing most of the casing to remain open. The ground remains in place as the casing is advanced, swallowing the spoils in the casing. The cutting ring has a smaller section area than the full-face area, which significantly reduces the influence zone of ground compaction around the casing. Advantages of open-face pipe ramming include the ability to swallow cobbles and boulders during the ramming process, that larger casings may be placed without the same degree of damage-zone influence, and that larger casings may be advanced with less energy as the cutting resistance from the leading edge is reduced to just the cutting shoe area or casing wall thickness. Cobbles and boulders as large as the inner diameter of the casing can be swallowed. This technique is typically used for casings larger than 6 in. (150 mm) in diameter.

Open-ended pipe ramming cutting rings are weaker than closed-face solid tips; thus the rings are more susceptible to crushing, erosion of material, or buckling and other damage during driving than with closed-face tooling. To strengthen the cutting ring, the leading edge of the first casing is usually reinforced by welding a steel band 12 to 18 in. (305 to 455 mm) wide around the exterior face of the casing. The banding provides two advantages: (1) it reinforces the leading edge, and (2) it decreases the frictional drag around the casing by creating an overcut. A steel band can also be welded around the inside edge of the leading section of casing. This internal band adds further reinforcement, protects the casing end from damage, guides any cobbles or boulders into the casing for removal when the leading edges are staggered to create a taper inward, and creates a clearance for the spoils within the casing. This clearance helps during the clean-out process and reduces the frictional drag within the casing. The leading edge of the cutting shoe is often beveled toward the inside of the casing to facilitate ground entering the casing, which reduces required drive energy. (For more information on cutting shoes, see Section 2.3.6.) For small-diameter casings, after the casing installation is complete, the ground that has entered the casing can be removed by applying compressed air or water from either end or the use of an auger boring machine. For larger diameter installations, an auger boring machine, skid steers, or HDD rigs with post hole augers can be used to mechanically remove the spoils from inside the casing. Cobbles and boulders may be removed by hand or pulled out with a cable and anchor bolt placed into the cobble or boulder.

For either the closed-face or open-face process, bentonite or polymer lubricants may be added to reduce the friction forces to allow installation of large-diameter casings with longer installed lengths or to reduce soil adhesion to the casing in some ground conditions, such as stiff clays or sands. A small-diameter steel supply pipe installed on top of the steel casing delivers the lubricant pumped by a small positive displacement pump. The start of the lubricant supply pipe should be installed approximately 24 in. (610 mm) behind the leading edge and protected from the ground by placing a circumferentially oriented steel plate in front of the discharge to allow side discharge and to distribute the discharge around the full diameter of the casing. Longer drives may require several sets of lubricant pipes. The lubricant may consist of water, bentonite, or polymer lubricants, as required for the ground conditions. Most lubricants are marketed by drill fluid component manufacturers, which are also available to provide technical guidance for the selection and use of their products for specific project ground conditions. The lubricants may be placed outside the casing, inside the casing, or both to aid in spoils removal, to reduce exterior or interior skin friction, and to maintain the integrity of the hole being created.

2.3 EQUIPMENT SELECTION AND SETUP

2.3.1 Guide Rails and Setting Grade

Typically, setups are positioned on the design line and grade. The accuracy of the installation is highly dependent on the initial casing alignment and stiffness of the casing. For this reason, it is imperative that the casing and hammer have stable support and remain precisely aligned throughout the ramming process. Consider the pipe ram installation as a marksman shooting at a target. Hitting the target is a function of the initial preparation and alignment of the equipment. Once launched, there is little ability to redirect the system. A stable foundation at the launch shaft is especially important for long horizontal installations to maintain their alignment.

Pipe ramming launch shafts do not require the use of reinforced back abutments, thrust blocks, or structural base slabs as is typically required for shafts used for jacking. Shaft sizes are directly tied to the length and diameter of the casing sections that can be used plus the length of the hammer and connecting devices for running the hammer. The launch shaft preparation will, in part, be dictated by jobsite conditions and contractor's equipment and will vary by project.

A solid foundation is critical to a project's success. Ground with insufficient bearing capacity can be displaced, resulting in foundation settlements or other movements. Foundation stability is a function of ground type, depth of excavation, and groundwater influence. The ground beneath the stable support system should be assessed for stability, such as liquefaction during hammer operation or bearing. Stabilization of the foundation may be necessary before or during construction of the shaft. Groundwater in granular soils within about 3 ft (1 m) of the launch shaft floor needs to be assessed for stability under cyclic loading conditions and mitigated. The launch shaft floor should be a poured concrete slab to grade or a compacted base comprising gravel covered with a steel plate or rig mats set to grade.

Because pipe ramming, by definition, is a nonsteerable means of casing installation, it is important to establish a grade tolerance at the desired casing maximum or minimum elevations. In challenging ground conditions, large grade variations may occur because of the ground conditions, but these grade changes should not be considered as a given and can be controlled to a limited extent by precise initial setup of the guide rails. Increasing the casing diameter may also increase accuracy while allowing for greater tolerances.

Casing support tracks can be a horizontal auger boring track system or a contractor-built system using I-beams, C-channels, or sheet piles (see Figures 2-2 through 2-4). For large casing sizes, two I-beams have been used to support the increased weight (Figures 2-5 and 2-6). Hydraulic casing leveling systems are also available from the hammer manufacturers. Installation of smaller diameter casing can often be accomplished without any type of support track, providing the support system is adequate to maintain the setup line and grade. Because auger boring is often used for cleaning the casing, the auger boring guide rails are often used for guidance.

2.3.2 Minimum Space Requirements

Shaft size, portal size, and placement of equipment should consider casing lengths, hammer length, and hose connections. Typical hammer

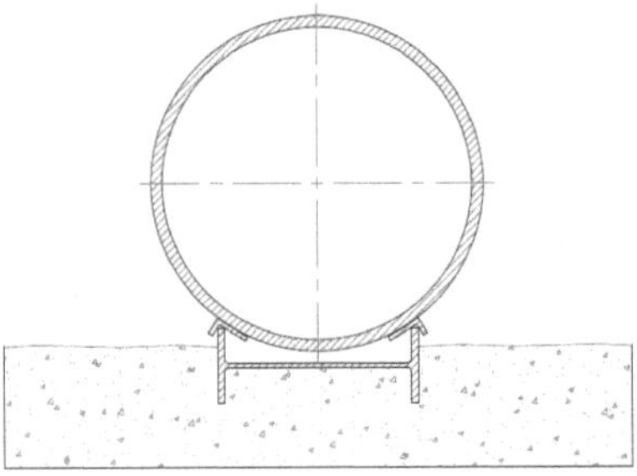

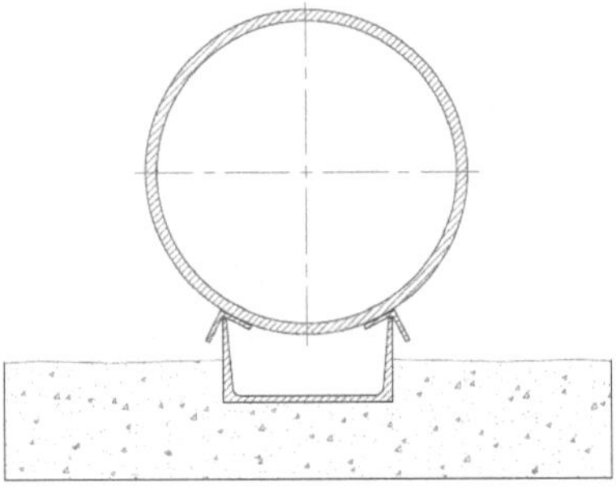

Figure 2-2. Alignment on I-beam (left).
Figure 2-3. Alignment on C-channel (right).

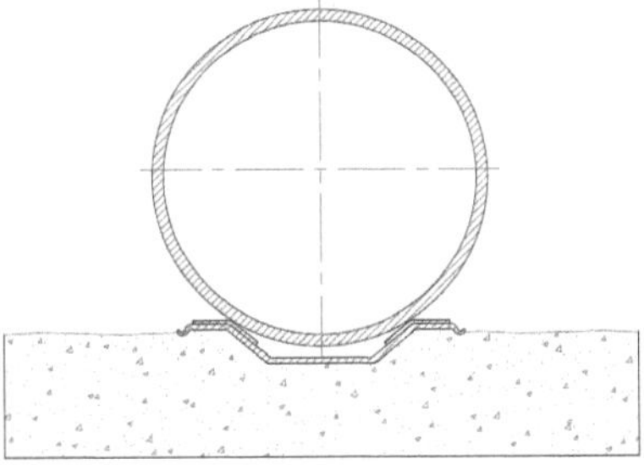

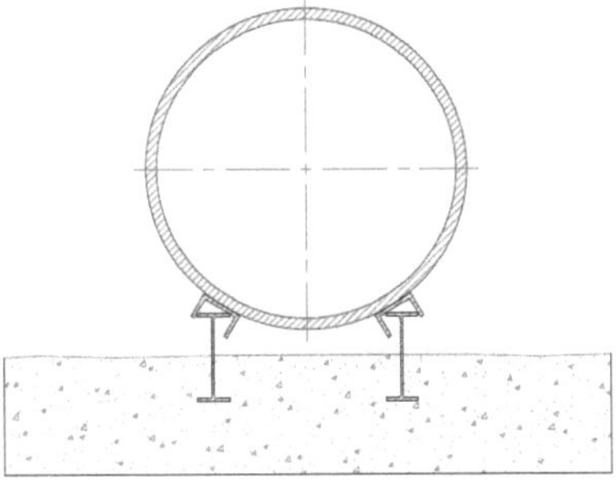

Figure 2-4. Alignment on sheet pile (left).
Figure 2-5. Alignment on two I-beams (right).

lengths are provided in Figures 2-7 through 2-9 from three different hammer manufacturers. The manufacturers' websites (see figure captions) provide current information on the hammer sizes.

The shaft or portal should account for the length of casing, usually 20 ft (6.1 m) or 40 ft (12.2 m), plus the hammer length and working space for welding and other tasks. Typically, a 3 to 6 ft (0.9 to1.8 m) gap between the back of the hammer and the shaft inside face is provided for assembly of the system and pipe alignment, plus 3 to 6 ft (0.9 to1.8 m) of clear space on either side of the casing to permit pipe alignment, welding room, or room to make other types of connections.

Working space around the top of the shaft or portal is needed for storing casing and for locating a crane, boom truck, excavator with arm, or hoist. The crane, boom truck, excavator with arm, or hoist is used for placing casing on the guide rails and positioning/attaching the hammer to the casing.

Figure 2-6. A typical pipe ramming setup with guide rails, casing, collet, drive cone, hammer, and connecting chains/straps on two I-beams.
Source: Courtesy of Hammerhead Trenchless.

Deeper shafts often require internal bracing. The location of the cross bracing should be designed such that (1) the hammer can easily be removed at the completion of one section of casing installation, (2) a full-length casing piece is placed, and (3) the hammer is then placed at the end of the next section of casing.

2.3.3 Depth

The excavation depth of a shaft should account for the thickness of the rail or other guide system, sumps, and drainage materials if used. Dry space should be provided directly under the casing for welding, typically 1.5 ft (0.5 m).

It is preferred to perform pipe ramming downslope. Where groundwater is likely or expected, ramming upslope should be considered to allow groundwater to drain toward the lower shaft. Should groundwater intrusion be expected, and for sloped shaft-to-shaft installations, consideration should be given for depth and area to install drainage systems to handle potential groundwater intrusion.

For installations that cross under buried structures or utilities, the depth of cover between the new casing and the existing buried structure or utility is an important consideration in both crossing design and guide system elevation. Various regulatory authorities, owners, and state utility locating laws require a specific clearance distance, which needs to be considered in selecting the depth of the installation. Additional considerations for determination of clearance distance from existing utilities are understanding the

MODEL	TOOL DIAMETER (IN.)	LENGTH (IN.)	WEIGHT (LBS.)	STROKES /MINUTE	AIR CONS. (CFM)
Mini-Atlas	5	36	132	580	60
Titan	5.7	61	302	285	141
Mini-Olympus	7	43	385	500	124
Olympus	7	66	496	305	177
Hercules	8.5	75	811	340	282
Mini-Gigant	10	48	1,014	430	353
Gigant	10	79	1,356	310	424
Koloss	14	92	2,601	220	706
Goliath	18	112	5,434	180	1,236
Taurus	24	144	10,580	180	1,766
Apollo	32	174	25,000	180	3,500

Steel pipe ramming machine GRUNDORAM

GRUNDORAM	DAVID	ATLAS	TITAN	OLYMP	HERKULES	GIGANT	KOLOSS	GOLIATH	TAURUS	APOLLO	Mini-Atlas	Mini-Olymp	Mini-Gigant
Machine-Ø (mm)	95	130	145	180	216	270	350	460	600	800	125	180	270
Rear cone Ø (mm)	112	145	160	195	235	300	400	510	670	900	140	230	330
Length (mm)	1490	1453	1545	1690	1913	2010	2341	2852	3645	4400	946	1080	1230
Pipe-in-pipe (mm)	-	-	-	-	-	-	-	-	-	-	250	450	450
Weight (kg)	59	95	137	230	368	615	1180	2465	4800	11500	60	175	460
Air consumption (m^3/min)	1,2	2,7	4,0	4,5	6,5	12,0	20,0	35,0	50,0	100	1,7	3,5	10,0
No. of strokes (min^{-1})	345	320	310	280	340	310	220	180	180	180	580	500	430
Impact energy (Nm)	230	420	800	890	1440	2860	6820	11600	18600	40500	180	720	2000
from pipe OD on	50	50	100	100	120	200	280	380	380	600	50	100	200

* = depending on soil type PK = short version Subject to change! 01.01/2004 E 1 Operating pressure 6-7 bar

Figure 2-7. Summary of TT Technologies hammers.
Source: Hammerhead Trenchless, https://www.hammerheadtrenchless.com/ (2019).

method of the original installation (i.e., open-cut versus trenchless and shored versus sloped excavation) and whether bedding material is present (and if the bedding material is saturated or dry, which will affect bedding and adjacent ground stability). It is also important to know the location of the utility relative to the launching and receiving locations. If the option exists, it is best practice to place the launching shaft closer to the utility being crossed because pipe ramming is an unguided method and the amount of clearance is best established at the start of the drive.

HammerHead® Trenchless Pneumatic Pipe Ramming Equipment Specifications HAMMERHEAD®

MODEL	4.00" RH SR	5.125" SR	5.50" SR BU	5.75" SR	7.00" AR	8.00" AR	12.00" AR	16.00" AR	20.00" AR	26XPR	34.00" AR
DIAMETER - IN (MM)	3.88 (98)	5.25 (133)	5.50 (140)	5.88 (149)	7.00 (180)	8.00 (200)	12.00 (300)	16.00 (400)	20.00 (500)	25.60 (650)	34 (860)
LENGTH IN (CM)	63.95 (162.4)	65.92 (167.4)	63.95 (162.4)	79.5 (201.9)	81.0 (205.7)	105.0 (266.7)	91 (231.4)	97 (245.1)	108.4 (275.3)	138 (351)	166 (422)
WEIGHT LB (KG)	136 (62)	214 (97)	170 (77)	305 (138)	511 (232)	900 (408)	1,568 (711)	2,701 (1,225)	5,750 (2,608)	12,500 (5,670)	27,500 (12,474)
AIR CFM (L/MIN)	68 (1,926)	98 (2,775)	68 (1,926)	132 (3,738)	235 (6,654)	308 (8,722)	600 (17.0)	1,050 (29.7)	1,290 (36.5)	3,000 (84,950)	4,800 (135,921)
TAILBOLT ACCESS	6	8	6	8	8	12	16	20	24	32	34
PRESSURE PSI (BAR)	110 (7.6)	110 (7.6)	110 (7.6)	110 (7.6)	110 (7.6)	110 (7.6)	110 (7.6)	110 (7.6)	110 (7.6)	110 (7.6)	110 (7.6)
REVERSE	7 turns	7 turns	7 turns	9 turns	Air Reverse	Air Reverse	Air Reverse	Air Reverse	Air Reverse	Air Reverse	Air Reverse
BLOWS PER MINUTE	343	333	317	295	302	196	236	232	177	162	120
ENERGY PER IMPACT (IN*LB/BLOW)	2132	3769	2132	4620	8745	17379	35384	60351	108703	192904	428254
CALCULATED IMPACT FORCE AT 1FT/MIN(TONS)**	30.5	52.3	28.2	56.8	110	141.9	347.9	583.4	801.7	1302.1	2141.3
REC. PIPE RAM SIZE IN (CM) *	2-4 (5-10)	4-8 (10-20)	2-4 (5-10)	6-12 (10-30)	8-20 (15-50)	8-24 (20-60)	12-30 (30-75)	16-48 (30-120)	20-54 (50-140)	24–84 (61–213)	36–180 (90–450)
REC. HDD ASSIST DRILL CLASS	N/A	N/A	N/A	N/A	24K	24-50K	33-80K	80K	80-150K	150K and larger	450 K and larger

RH-Replaceable Head, SR-Screw Reverse, AR-Air Reverse.

**Numbers shown are calculated based on 110psi and assuming

Figure 2-8. Summary of Hammerhead Trenchless hammers.
Source: Hammerhead Trenchless, https://www.hammerheadtrenchless.com/ (2019).

IHC HYDROHAMMER®
TECHNICAL DATA IMPERIAL

HAMMER S SERIES		S-30	S-40	S-70	S-90	S-120	S-150	S-200	S-280	S-350	S-500	S-600
Max. blow energy	lb.ft	22,127	29,503	51,630	66,381	88,508	110,635	147,513	206,518	258,147	368,783	442,539
Min. blow energy*	lb.ft	2,213	2,950	5,163	6,638	8,885	11,063	14,751	20,652	28,765	36,848	44,254
Blowrate	bl/min	65	65	50	46	44	44	45	45	40	45	44
Ram weight	lb	3,527	4,850	7,716	9,921	13,668	16,534	22,046	29,982	36,000	55,115	66,138
Hammer weight	lb	8,599	10,362	18,298	21,348	31,526	35,714	56,878	67,240	72,000	126,764	141,093
Length of hammer	ft	20.00	22.47	24.28	26.43	26.79	29.2	29.84	34.09	37,35	39.18	41.72
Oil flow	g/min	46	46	66	66	122	122	211	211	211	423	476
Power pack type		P 175	P 175	P 250	P 250	P 460	P 460	P 700 / P 800	P 700 / P 800	P 700 / P 800	O.R.	O.R.

HAMMER S SERIES		S-800	S-1200	S-1400	S-1800	S-2000	S-2500	S-3000	S-4000
Max. blow energy	lb.ft	590,056	885,079	1032,952	1327,618	1475,131	1843,905	2212,686	2950,248
Min. blow energy*	lb.ft	59,005	88,508	103,259	132,762	147,513	199,141	281,748	292,812
Blowrate	bl/min	45	40	35	35	35	32	35	36
Ram weight	lb	88,183	132,275	152,116	198,413	220,459	275,577	330,693	440,924
Hammer weight	lb	182,981	308,642	326,279	454,145	487,213	573,201	639,340	947,987
Length of hammer	ft	47.93	46.91	52,79	54.17	56.87	62.42	68.09	66.96
Oil flow	g/min	581	581	871	871	1.162	1.162	1.347	1.744
Power pack type		O.R.	O.R.	O.R.	O.R.	O.R.	O.R.	O.R.	O.R.

HAMMER SC SERIES		SC-110	SC-150
Max. blow energy	lb.ft	81,132	110,635
Min. blow energy*	lb.ft	8,113	11,063
Blowrate	bl/min	45	45
Ram weight	lb	17,416	24,250
Hammer weight	lb	31,085	41,226
Length of hammer	ft	18.88	21.75
Oil flow	g/min	122	122
Power pack type		P 460	P 460

POWER PACK TYPE		P-250	P-460	P-700	P-800
Max. pressure	psi	5,076	4,786	4,931	4,641
Max. oil flow	g/min	66	122	184	211
Power	hp	228	526	700	768
Length	ft	11.48	13.22	15.91	14.76
Width	ft	5.05	5.07	6.23	5.91
Height	ft	6.71	7.30	7.69	8.27
Net. Weight	lb	7,716	11,905	14,550	17,196
Weight incl. fuel and oil	lb	9,921	16,314	20,945	21,385

**The standard minimal energy setting is about 10% of the hammer's maximum energy.*
When using the high frequency/low energy mode, the energy can be reduced to a minimum of 2% to 5%.

Figure 2-9. Summary of IHC Hydrohammer hammers.
Source: Courtesy of IHC IQIP.

2.3.4 Hammer Selection

Hammer selection is best left up to the contractor and hammer manufacturer. Specifications should typically only set the minimum tooling required for installation. Hammer manufacturers have readily available tables of their specific hammer models that indicate energy ratings and typical ranges of casings installed with that hammer (Figures 2-7 through 2-9).

2.3.5 Air Compressor

Pneumatic hammers require an air compressor to supply compressed air to advance an internal piston that strikes on the internal anvil to generate the driving energy. It is important that the air compressor, whether new or used, maintain the needed air volume and pressure to operate the pneumatic hammer properly.

Compressor discharge air temperature should be monitored to prevent damage to hammer components. Older compressors may produce higher discharge air temperatures, which will damage the internal hammer components over time. Proper operating air temperature conditions should be discussed with the hammer manufacturer to determine appropriate air supply operating temperature ranges.

Air compressor output pressure and volumes are rated at sea level conditions. Compressor efficiencies can vary widely because site factors typically differ from sea level conditions. To mitigate this efficiency issue, plan on having a compressor capable of providing 25% more volume than required to run the pneumatic hammer after allowing for elevation and other performance factors. Air compressors will see an approximately 5% loss in efficiency with every 1,000 ft (305 m) of elevation increase. For example, an air compressor operating in Denver, Colorado, will be 25% to 30% less efficient than the same air compressor operating at sea level. A low air pressure will not stop a hammer, but the hammer energy available for driving will be reduced, and the hammer may not have enough energy to drive the casing into the ground.

Air supply hose length and diameter provide resistance to the air supply delivered from the compressor. Typically, a 2.5 in. (65 mm) diameter supply hose is used to deliver compressed air to the pneumatic hammer. This size of air hose allows for less pressure loss over short distance equipment configurations. It is important that the air system has the properly sized fittings and connections between the air compressor, lubricator, and pneumatic hammer. Restrictions in the air flow from poorly designed supply hose diameter and length combinations will cause friction and heat buildup, reducing hammer efficiency. Air hose diameter and length should be assessed for flow capacity for each installation for optimal pipe ramming operation. There is an approximately 3 psi (20.7 kPa) drop in pressure in a 2.5 in. (65 mm) diameter supply hose per 100 ft (30.5 m) of air hose.

Air compressor efficiency can be evaluated by counting the number of hammer strokes per minute at full volume and pressure, and then comparing those numbers with the manufacturer's literature. The number of piston strikes per minute varies by tool size and by manufacturer. To mitigate poor performance, a second air compressor with a parallel delivery system can be used. The two air delivery systems are joined at the tail of the hammer.

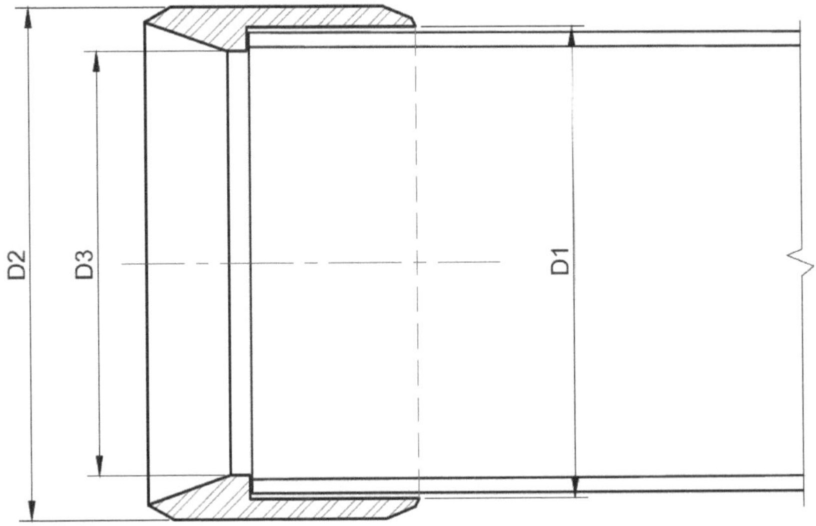

Figure 2-10. Example of a premanufactured cutting shoe.

2.3.6 Cutting Shoe

A cutting shoe should be added to the leading edge of the casing and is an integral part of the pipe ramming process (Figure 2-10). Before pipe ramming begins, the cutting shoe is fully welded to the leading edge of the casing. Its primary purposes are to

- Stiffen the leading edge of the casing against crushing and longitudinal, radial, and circumferential buckling through difficult ground conditions;
- Create an overcut and undercut, which can reduce both external and internal casing friction during casing advancement;
- Help to protect the casing's coating;
- Create an annular gap for the flow of a bentonite/polymer lubricant;
- Create a chamfered/beveled leading edge to help direct spoils into the casing; and
- Influence the casing alignment accuracy by deflecting buried objects away from the leading edge.

Cutting shoe design and fabrication are influenced by local contractor experience. Based on contractor, manufacturer, and engineering experience, the external cutting shoe band should extend 270 degrees to 360 degrees circumferentially around the outside of the casing depending on ground conditions. The internal cutting shoe band usually extends completely (360 degrees) around the inside of the casing. The bevel on the external and internal bands should direct the flow of spoils inward. An outward bevel cutting shoe increases the chance of densifying the ground conditions and limiting the drive length. If dense ground conditions are encountered, more power will be required to displace the ground at the leading edge of the casing. If a cobble or boulder is encountered, the chance of casing failure by buckling is increased with an outward beveled cutting shoe design. Cutting shoes should be designed and installed by an experienced contractor because length, position, and amount of coverage around the casing inside and outside diameter will influence the amount of energy necessary to drive the casing.

Premanufactured cutting shoes are typically used for short bores in soft soils. They provide limited stability, are typically made from a mild steel, and do not include provisions for lubrication along the casing walls. As shown in Figure 2-10, Dimension D1 is selected to be slightly larger than the outside diameter of the casing so the shoe can slip over the casing. The cutting shoe is welded to the casing so it remains in place and attached during the pipe ramming process. Dimension D2 is selected to create an overcut and to allow the injection of lubrication to the outside surface of the casing. Dimension D3 is selected to create an undercut to help reduce friction forces from the ground inside of the casing.

Homemade or welded on-site cutting shoes are better because of the contractor's ability to design the shoe based on the anticipated ground conditions, the ability to use a higher-grade steel (Figure 2-11), and the

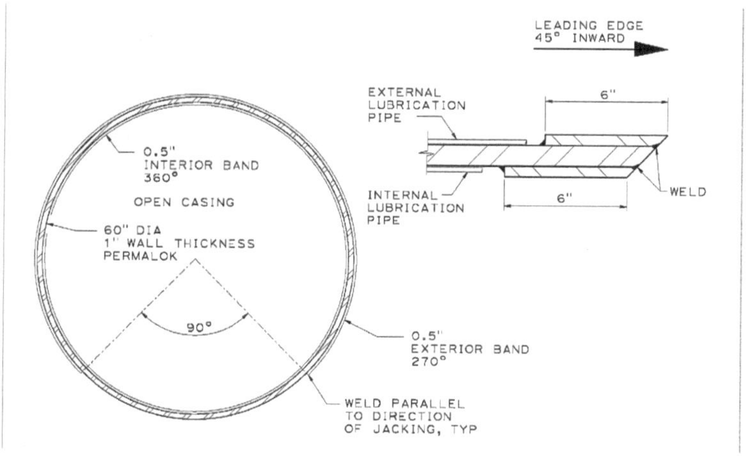

Figure 2-11. Example of a homemade cutting shoe with internal and external banding and welded with a chamfered/beveled leading edge.
Source: Courtesy of McMillen Jacobs Associates.

ability to include lubrication that can be distributed around the full outside circumference of the casing if required.

Lubrication is always an advantage in a pipe ramming operation and is typically dispensed behind the cutting shoe. Lubricant usage helps counteract ground friction inside and outside the casing. The lubricant also aids in the removal of spoils from within the casing after the ram is completed. Lubrication should start when the ends of the lubrication pipes start to become buried in the ground and continue until the end of the drive. If lubrication is anticipated on a project, start the injection during the initial casing drive. Applying lubrication after a casing has become stuck is often not successful because the lubrication cannot reach the areas of high friction.

In cobble and boulder ground conditions, contractors have hard-faced the leading edge or welded hard metal or carbide teeth to the front of the cutting shoe. The teeth both loosen hard ground and point load any cobbles and boulders encountered by the leading edge, and they also help to fracture rock pieces if they are directly in the path of the casing (Figure 2-12). Roller cutters are ineffective in pipe ramming because there is no rotation of the leading edge of the casing to engage the roll cutting tools.

Figure 2-12. Different kinds of cutting shoes used for pipe ramming. Source: Courtesy of Staheli Trenchless Consultants.

A sonde can be added to the front of the casing to locate vertical and horizontal alignments. If the sonde is added, it is done by welding a pipe to the outside of the casing, cutting slots, and filling the slots with epoxy so the sonde can transmit its signal past the casing. The line can be used as the lubrication line while ramming because ramming cannot occur with the sonde in or on the casing. In general, there is no steering component to a pipe ram installation. Occasionally, if an installation is losing grade, contractors have reported removing all spoils from inside the casing and removing an obstruction or shimming the bottom of the leading edge to try and deflect the casing. This attempt at grade correction is not considered active steering but rather a contingency plan for the loss of grade. An alternative remediation to recovering a ram project is to ram a larger diameter casing from the receiving shaft and attempt to engulf the original casing from the other direction.

2.3.7 Casing Selection and Wall Thickness

Casings must have enough strength and life span to handle both long-term loading conditions and short-term loads during construction. These requirements are factored into the design of casing type and wall thickness. For pipe ramming, the highest loading conditions typically occur during construction.

When selecting a casing type, size, and wall thickness, some of the critical variables to consider are the ground conditions, line and grade requirements, length of the ram, site layout, drive hammer energy, and transmission of hammer energy to the cutting shoe. The casing must be of adequate size and wall thickness to withstand the earth face resistance and casing surface frictional forces for the desired overall ram length. Many hammer manufacturers have published minimum wall thicknesses for various casing diameters on their websites. Table 2-1 presents recommended minimum wall thicknesses for various steel casing diameters. Based on long-term train loadings, the American Railway Engineering and Maintenance-of-Way Association (AREMA) may require different wall thicknesses than those shown in Table 2-1. Consult the AREMA *Guidelines for Railroad Grade Separation Projects* (2019) for minimum casing wall thickness for crossings under railroad rights-of-way.

Both casing and cutting shoe strength typically increase as the complexity and challenges of the project increases. The longer the bore or the more complex and difficult the ground conditions, the higher the required casing strength. Additional casing strength is required to withstand the higher hammer applied force necessary to overcome the higher casing friction resistance from a longer drive and the higher ground resistance force applied to the cutting shoe by adverse ground conditions. Casing strength is necessary to resist plastic deformation of the casing geometric shape and

Table 2-1. Casing Wall Thickness Chart.

Recommended wall thickness for pipe ramming		
Pipe size [in. (mm)]	Min wall thickness [in. (mm)]	Dimension ratio (DR)
4 (100)	0.188 (4.8)	21
6 (150)	0.188 (4.8)	32
8 (200)	0.25 (6.4)[a]	32
10 (250)	0.25 (6.4)[a]	40
12 (300)	0.25 (6.4)[a]	48
14 (350)	0.312 (7.9)[a]	45
16 (400)	0.312 (7.9)[a]	51
18 (450)	0.312 (7.9)	58
20 (500)	0.375 (9.5)[a]	53
22 (550)	0.375 (9.5)[a]	59
24 (600)	0.375 (9.5)	64
26 (660)	0.500 (12.7)[a]	52
28 (710)	0.500 (12.7)[a]	56
30 (750)	0.625 (15.9)[a]	48
32 (815)	0.625 (15.9)[a]	51
36 (900)	0.625 (15.9)[a]	58
42 (1,050)	0.75 (19.1)[a]	56
48 (1,200)	0.75 (19.1)[a]	64
54 (1,350)	0.75 (19.1)[b]	72
60 (1,500)	0.75 (19.1)[b]	80
66 (1,650)	0.75 (19.1)[b]	88
72 (1,800)	0.75 (19.1)[b]	96
78 (1,980)	1.00 (25.4)	78
84 (2,100)	1.00 (25.4)	84
96 (2,400)	1.00 (25.4)	96
108 (2,700)	1.25 (31.8)	86
120 (3,000)	1.25 (31.8)	96
144 (3,600)	1.25 (31.8)	115

Note: Minimum wall thickness recommendations should be verified and designed/modified for specific ground and project conditions.
[a]Recommended minimum wall thickness is greater than required in AREMA (2019).
[b]Check AREMA (2019). AREMA requires a thicker-walled casing for E80 loading.

casing material. It is a function of the casing material properties, mainly the casing material strength, and the thickness of the casing or cutting shoe material. In general, for steel casing a minimum material strength of 36 ksi (248.2 MPa) is recommended. Note that high-strength steel is acceptable but often comes with reduced ductility. This means that the high-strength casings, although producing a thinner wall casing, typically crack (fail) at lower deflections than a low-strength steel.

The casing diameter and wall thickness have a direct correlation to the length and size of the ram, the energy of the drive hammer, and the amount of energy that can be transmitted along the casing to the leading edge to advance the casing. The casing thickness should be designed both to transmit enough energy to the leading edge to advance the casing and to withstand the drive energy of the hammer without buckling.

Casing coating if used for a pipe ramming operation will likely be damaged by the ramming process, thus, the slick bore process is needed. Some contractors use a thicker coating to counteract any frictional damage during installation. If a coating is used, consult with a coating manufacturer for written recommendations. If steel corrosion is a concern, consult a corrosion specialist and consider installing a casing and then installing the steel carrier pipe with the coating and lining needed inside of the casing. It is typical to use nonconductive spacers to prevent electrical contact between the casing and carrier pipe if both are subject to corrosion. The annulus may then be grouted for further corrosion protection and to reduce the future potential for settlement of the overlying ground.

2.3.8 Hammer Attachment

There are different ways to connect a hammer to the casing. The most common way of connecting a hammer to a smaller diameter [less than 60 in. (1,525 mm)] casing is to use collets or segments (Figure 2-13). Collets/segments come in three or four pieces with a taper lock connecting the hammer to the collets/segments to the casing. They come in pieces to assist in shipping the collets because of their weight and size. For casings larger than 72 in. (1,830 mm) in diameter, a combination of drive cones and collets are used. A drive cone is a forged cone that transitions from the collets to the large-diameter casing. Efficient drive cones should be designed for high efficiency for transmitting dynamic loads. Typically, a drive cone should be a minimum of 10 times stiffer than the casing. Wall thickness of the casing is important for transmitting energy to the cutting shoe but is also necessary to lock the collets and drive cones into the casing without the internal force applied by the collets splitting the casing. Collets are designed and forged for the casing sizes recommended by the manufacturer for a specific casing diameter. Too thin of a wall could result in the collets not locking into the casing, potentially causing damage to it during the

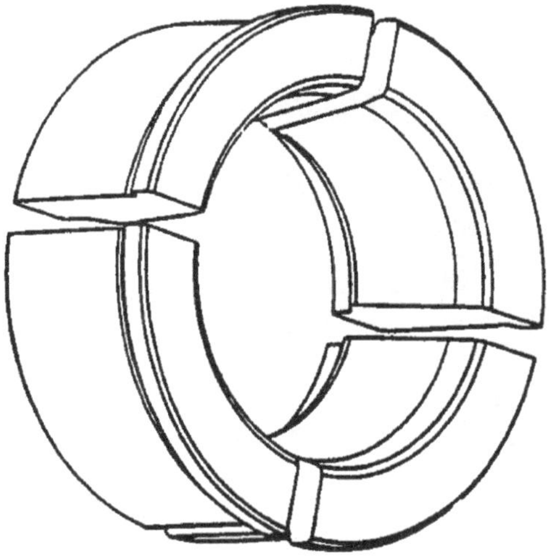

Figure 2-13. Typical configuration for a multipiece collet to protect the end of the casing less than 60 in. (1,525 mm) from flaring.

ramming process including splitting the end of the casing. A wall that is too thick could result in the collets not fitting into the casing. For this reason, the casing dimensions need to be carefully coordinated with the hammer and collet manufacturers.

A supporting guide system is typically used for casing placement. The guide system includes either tracks or rails both to support the casing and hammer during the assembly and drive process and to align the casing and hammer system with the design line and grade. The supporting guide system with hammer and casing are set to grade prior to starting the pipe ramming process. Hydraulic or mechanical leveling jacks have been used to support the guide system, and in many cases these jacks can save time in setting a design grade. After the first casing is set on the tracks or with leveling jacks, the collets are then loaded into the casing. The nose of the hammer is inserted into the opening of the collets. If drive cones are used, the drive cone is inserted into the casing and the collets are pulled into the drive cone (Figure 2-14). In most pipe ramming applications, the back end of the hammer is suspended or supported on a slide and locked with a cinch assembly into the collets and casing (Figure 2-15). A cinch kit is used to keep the hammer tight and connected to the collets and casing to obtain more efficient energy transfer between hammer and collet. A hammer cradle can be used to support the hammer in larger diameter ramming applications. The cradle supports the weight of the hammer and allows the hammer to be more easily and quickly centered into the collets and casing.

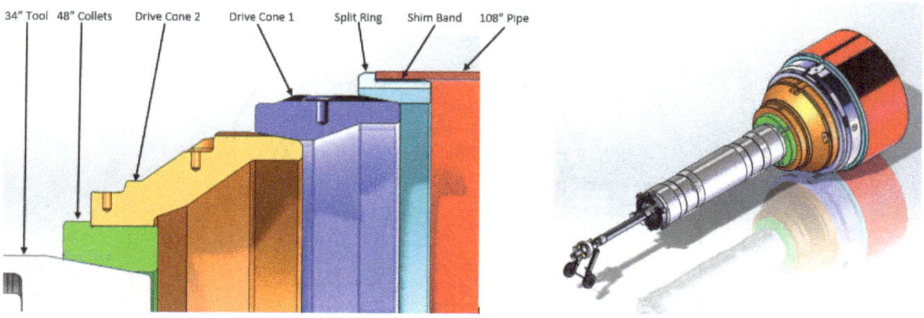

Figure 2-14. Typical configuration for a collet-and-cone assembly to protect the end of a larger diameter casing from flaring.
Source: Courtesy of Hammerhead Trenchless.

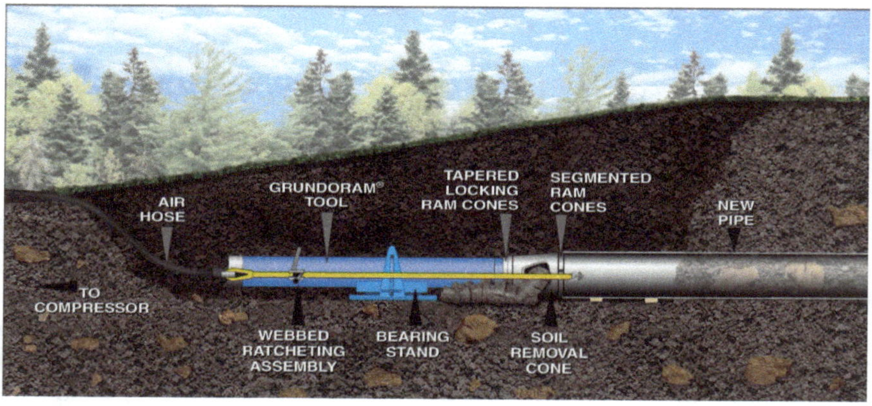

Figure 2-15. A typical pipe ramming setup with casing, collet, ram cone, hammer, bearing stand, and connecting straps.
Source: Courtesy of TT Technologies.

Several factors affect the available hammer energy and the energy transfer efficiency between the hammer and casing. The available hammer energy is a function of hammer size, the size of the energy supply (hydraulic or air compressor), and the angle of the hammer relative to the ground. A vertical hammer application provides a different available energy than a horizontally mounted hammer. The efficiency of the system is discussed in Section 2.3.9. Presently, the efficiency of pipe ram systems is not consistent, mainly because there is a wide range of stiffnesses for connections between the hammer and the casing. With a stiff connection, the efficiency will be higher. Straps, chains, or a cinch kit are used to cinch the hammer tight to the collets, thereby increasing the connection stiffness (Figure 2-6). When the hammer is not locked into the collets, a swimming motion or slapping can result, reducing the longitudinal energy transfer to the casing.

As the casing advances, the spoils enter the casing at an equal pace. With the vibrations and energy transfer during the ramming process, the spoils inside of the casing begin to bulk. Two holes are typically included in the drive cone to allow the expansion of spoils to exit the casing. In some cases, a soil port has been used to allow the spoils to come out of the back end of the casing as the casing is advanced. This soil port can help in limiting the amount of spoils left to be removed. A soil port is not conducive to the removal of cobbles and boulders.

When selecting support equipment, a contractor should consider the hammer weight, drive cone weight, and the reach of the track hoe or backhoe. A crane will be required with larger hammers and drive cones. The weights of the hammers and drive cones can be significant. See Figures 2-7 through 2-9 for the weights of available hammers.

2.3.9 Efficiency of Casing Penetration or Advancement

Individual casing pieces are assembled into a casing string for pipe ramming. The components of a pipe ramming system are hammer (energy source), collet (couples hammer to casing), and casing joints (weld or mechanical couplings). The energy transferred to the cutting shoe from the hammer is the system efficiency. The system efficiency is the actual applied hammer energy minus the energy loss in the hammer to collet connection; in the collet to casing connection, at each joint, from the internal friction within the casing; and from friction between the casing and the surrounding ground. Pipe ramming can be done on any length of casing, provided the driving force transmits enough energy through the casing to the cutting shoe to penetrate the ground without crushing the casing. Any discontinuities between components attenuates energy, thus affecting the penetration rate during the driving process (Meskele and Stuedlein 2015). There are many publications on the dynamic driving of piles that provide a much more in-depth review of the subject. In general, the system efficiency and strength of the components will dictate the overall achievable drive length.

The penetration rate is defined as the amount of advance over a specific time interval. The rate of penetration of a casing from a specific hammer depends on the system efficiency. Resisting forces include friction between the surface area of the casing and the ground, and the strength of the ground being cut at the leading edge of the casing. The amount of driving force exceeding the resisting force is the amount of energy available to advance the casing. Each hammer blow advances the casing. Hence, the number of blows in a time interval multiplied by the advance from each blow provides the penetration rate. A higher available energy applied at a lower number of blows can produce the same penetration rate as a lower available energy blow applied at a higher number of blows. The typical penetration rate is measured in inches or feet per minute. It is the penetration rate on most pipe ramming projects that has the largest effect on project cost variation. Given the same casing and hammer size, ramming in clayey soil conditions is typically faster than ramming in cobble and boulder conditions.

Most pipe ramming projects consist of ramming 20 or 40 ft (6.1 or 12.2 m) casing pieces. An economical rate of penetration typically varies from 0.5 to 1.0 ft/min (0.15 m to 0.3 m/min). However, the system efficiency and actual ground conditions will determine the actual penetration rate. Common pipe ramming distances are less than 200 ft (61 m) and can range from 12 in. (305 mm) up to 60 in. (1,525 mm) in diameter. Diameters greater than 60 in. (1,525 mm) add a degree of difficulty.

A concern to address at the start of the pipe ram is recoil—the casing rebounding backward instead of forward. For the casing to advance in pipe ramming, friction is required between the outside of the casing and the surrounding ground. In more challenging ground conditions, such as loose cobble and boulders or with the pipe crushing method (Section 2.5.3), the casing does not have enough skin friction to advance. Therefore, the rebound or recoil is seen at the initial start of the ram. Often contractors use a pully or winch system to provide positive force on the first section of casing to reduce the effect of the rebound (Figure 2-16). As the casing continues to advance, the recoil or rebound is lessened. Some contractors have used hydraulic jacking systems to assist in minimizing the rebound.

The first section of casing installed is critical for setting line and grade. The line and grade of the casing must be checked at this time. Tracking the advancement and direction of each foot throughout the first section of casing offers the opportunity to try and change line and grade if needed. After the first section of casing has been installed, it becomes extremely difficult to change line and grade as the casing has now set a confined path to follow.

In addition to monitoring line and grade, it is important to monitor for any reduction in advancement per stroke and if refusal is reached. If refusal occurs before completing an installation, the contractor should assess whether the penetration rate slowly reduced then stopped (indicating the possibility of insufficient available energy for the drive) or if there was a sudden stop in penetration rate (indicating a possible buried object). A commonly used mitigation is to clean out the casing and see if a buried object is at the leading edge. Occasionally, the drive is close to the exit point. If that is the case, the receiving end may be excavated to expose the end of the casing, thereby completing the drive. Caution should be used when excavating the spoils within the casing when groundwater is present. If the installation is under the groundwater table, there may be requirements within the specifications to maintain an earth plug within the casing, which would preclude excavating to inspect the leading edge of the casing.

Figure 2-16. A pully and winch system used to minimize rebound.
Source: Courtesy of Staheli Trenchless Consultants.

Dewatering directly in front of the casing may be required to help with the inspection of a buried object. If cleaning out the casing and dewatering cannot be completed, excavation from the ground surface to explore the point of refusal may be the only safe recourse. In some cases, removing some of the spoils from the inside of the casing will reduce the total friction and weight on the casing, allowing a once stuck casing to advance.

When ramming in free-flowing sand, an earth plug needs to be created by installing sandbags, built up slightly behind the face of the casing. The sandbags are placed inside the casing in an effort to inhibit the free-flowing sand from entering and flowing into the casing.

While monitoring the penetration rate, it is also important to check for settlement or heave along the alignment. With free-flowing sands, a void can be created if an earth plug is not maintained in the casing. Heaving can occur when a buried object blocks the opening of the casing, forcing the ground up and out in front of the casing (also known as plowing). If a void is formed over the casing, remedial grouting should be done only after the casing is fully installed because the grout could set up around the casing and lock the casing in place.

There have been both vibration and noise monitoring studies completed on pipe ramming projects. Some jobsites are not suitable for pipe ramming because the noise from the impact force of the hammer on the casing and the sudden air release after each blow are too loud. For sites that are near sensitive receptors, mufflers can be added to a hammer assembly. Pipe ramming from within a shaft, depending on the depth, will buffer and redirect the noise upward in most cases.

Numerous vibration studies (Werner 2003) show no significant vibration effect on the tracks or ties underneath a railroad resulting from the pipe ramming process.

2.3.10 Lubrication

Lubrication pipe(s) can be placed on the outside of the casing and/or on the inside of the casing. In general, lubrication is only applied on the inside for diameters less than 24 in. (610 mm) where air is used to remove the spoils. Lubricants usually consist of a polymer (i.e., a slippery lubricant) to reduce the friction between the spoils and the casing. The lubricant also reduces friction when using a cleaning pig to eliminate the spoils.

Lubricant applied to the outside of the casing is intended to reduce the friction along the casing wall and ground interface. Such lubricants typically consist of either bentonite-based fluids or polymers.

Lubricant typically enters the 12 o'clock position of the casing and is dispersed just behind the cutting shoe, which creates an annulus for the lubricant. In unstable ground conditions, a short shroud may be placed over the lubrication discharge that extends around the casing circumference to allow full circumferential distribution of the lubricant.

An on-site mixer is used to blend water with polymers/bentonite to create the lubricant. Prepared lubricant is typically discharged into a storage container for ready application during a drive. Selection of an appropriate lubricant depends on the anticipated ground encountered along the pipe ram alignment. Distribution of the lubricant is provided by a pump and a hose connected to the lubrication pipe attached to the casing. The size and type of pipe used for conveying lubrication depend on whether the lubrication pipe is only used for applying lubricant or whether it is used for other purposes like hosting a tracking sonde. If it is only for lubrication, then the lubrication pipe on the casing typically consists of 0.5 in. (13 mm) diameter schedule 40 steel pipe(s) with threaded connections welded to the top of the casing at the 12 o'clock position (Figure 2-17). Threaded connections allow for faster installation as the casing is advanced. Angle iron can be used to protect the leading edge of the lubrication steel pipe discharge at the front end of the casing. When the intent is to lubricate both the outside and inside of the casing from the same lubrication pipe, a hole may be cut through the front end of the casing at the lubricant discharge location to disperse lubrication both to the outside and inside. Depending on the ground conditions, it may be prudent to protect the lubrication pipes with channel iron. Lubrication is usually applied along the entire length of the pipe ram. Multiple lubrication pipes can be added to dispense lubrication at different distances along the casing or at difference locations around the circumference of the casing on longer drives.

2.3.11 Joining Casing Pieces

Joining the casing pieces can be accomplished by either a mechanical joint or by welding. The joints are required to structurally connect the casing pieces such that the maximum amount of drive energy is transferred through the casing without damage to the joint. Following initial casing placement, casing assembly typically consists of lowering a second casing piece into the shaft and setting it on the guide rails. It is important to inspect both casing ends to ensure the ends are free of damage, are square and round geometry, and have the proper bevel before making the connection. Mechanical connections are assembled by following the manufacturers' recommended procedures. Welded connections should be only done by welders qualified in accordance with ANSI/AWS B2.1 (AWS 2014) and ANSI/AWS D1.1 (AWS 2020).

If the joints are connected by welding, apply a full penetration multipass butt weld: 100% penetration weld utilizing E6010 root pass and E7018 cover passes to fill the bevel. Factory midweld casings should be avoided unless the welds are guaranteed 100% penetration welds (Figure 2-18).

In addition to the butt weld, 2 × 6 in. (50 × 150 mm) straps are often welded horizontally over the butt welds. A full penetration weld is required on each of the passes. Many pipeline projects require certified welders and qualified welding procedures, because in many cases the casings are x-rayed for any weld imperfections. The welding process can take up to 4 hours to complete, depending on the casing diameter, the number of welders used, inspection and x-ray requirements, and coating repair.

A mechanical push-together, interlocking, nonwelded joint (such as Permalok or Tri-lock) has been used in several pipe ramming projects.

Figure 2-17. Typical configuration in which a smaller diameter steel pipe is welded to the top of the casing to allow dispersal of lubricant directly behind the cutting shoe. In addition, adding a small hole in the casing allows lubricant to enter inside of the casing.
Source: Courtesy of Hammerhead Trenchless.

These projects utilize the quick-locking joint precision machined to reduce the welding required in pipe ramming and give a precise, straight assembly when connected together. Historically, the push-together, non-welded joint has encountered some energy transfer issues because of pipe chatter. Permalok and other manufacturers are continuing to improve the interlocking joints specifically designed for pipe ramming applications to address these issues and improve the locked joint and transfer of energy. Even with the push-together joint, some contractors will elect to add some welding to ensure the casing ends remain engaged during the ramming process.

Whether the casing segments are welded or push-together, it is important that each casing segment is straight, round, uniform in circularity, and with square ends. The casing segment being added should be placed on the tracks/guide rails and checked to ensure the two joining casing segments mate up and are square. If the ends of the casings are not square, a bias can be introduced, which will result in a curved (horizontal, vertical, or compound) alignment. The welding process can also heat up the casing,

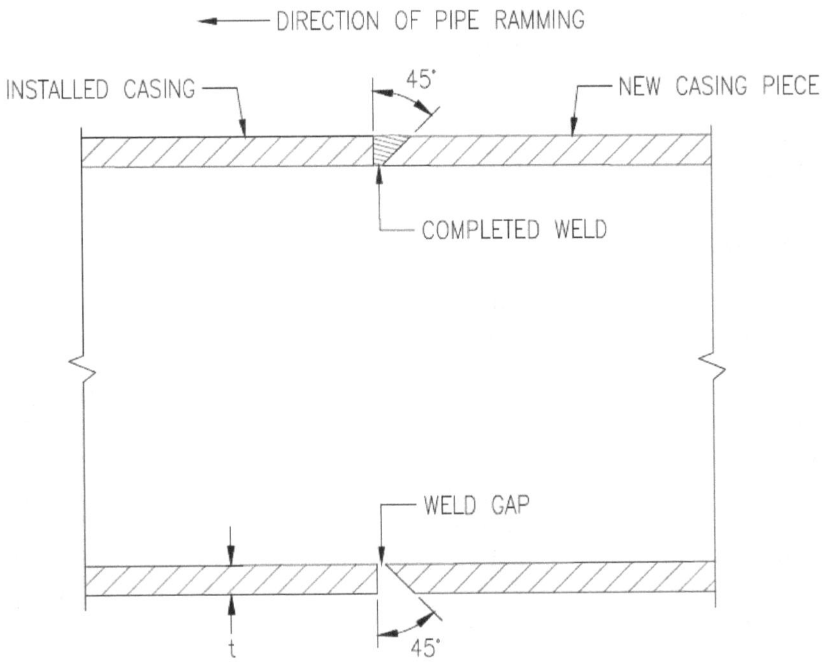

Figure 2-18. Welding detail for rammed casing.

and without proper care, a bias can be introduced into the pipe string. Uneven welding may cause the same problem.

The pipe ramming process introduces significant energy onto the end of the casing during advancement. The hammering effect, especially with thinner-walled casings, can damage the end of the casing by plastic deformation or buckling. Damaged ends will not align or mate with the next casing section. Introduction of a bias in the casing orientation needs to be assessed with each joining of the casing sections. If the end of a casing is damaged, the contractor will need to carefully cut off the damaged, out-of-round portion of the casing. The damaged portion of a steel casing will typically be cut with an oxyacetylene torch. The initial cut should be made using a mechanized pipe cutting machine that is set to cut vertically at the desired location to make a square end on the pipe (Figure 2-19). Next, the cutting machine is relocated, the torch is set at the proper bevel angle, and the bevel is cut. This procedure is followed by hand grinding for final preparation.

Freehand field cutting of the casing is an inaccurate process and may result in the casing end not being square. The contractor, welders, and the field inspectors must check and confirm end squareness and mating to eliminate a curved alignment. A major effort is needed to ensure a straight pipe ramming installation. If the casing is installed with a curve, ramming forces will increase and casing refusal can occur. In addition, the casing installation may be curved when completed. Note that intentionally curved pipe rammed crossings have been completed occasionally under waterways by joining casing pieces at an incremental design angle to create an arced path.

2.3.12 Spoil Removal

The spoils inside the casing can be cleaned out in a variety of ways and at any time. The most common way of cleaning out the spoils is using flighted augers with an auger boring rig or HDD drill with post hole auger. Water jetting, vacuum excavation, or blowing the spoils out using an air compressor are other options, depending on the ground conditions and the diameter of the casing.

In clayey, gravelly, and sandy conditions with casings less than 24 in. (610 mm) in diameter, contractors will typically try to blow out the spoils by using the air compressor. This method has proven to be ineffective in many cases, and a secondary piece of equipment might be needed to clean out the spoils. The most common method used is a flighted auger advanced with an auger boring rig. Many contractors have adapted their HDD drill rigs to clean out the spoils inside the casing. Vacuum excavators or water jetting trucks have also been used in removing the material inside of the casing. When a casing is large enough for personnel entry, a skid steer or mini excavator is used to remove the material. Appropriate safety measures for personnel entry into a confined space should be followed.

2.3.13 Carrier Pipe Installation

Whenever a carrier pipe is installed within a rammed casing (a two-pass construction method), the carrier pipe is set to line and grade using some type of support (e.g., skids, spacers, centralizers, blocking, racks). The supports need to be compatible with the carrier pipe material, corrosion requirements, and life expectancy of the project. For metallic carrier pipes, this may mean that the supports need to be manufactured from insulating material to electrically isolate the casing from the carrier pipe.

Supports have different uses and advantages. Racks are common when multiple large-diameter carrier pipes are installed in a casing. Casing centralizers are common in pressure lines, because they place the carrier pipe at the same vertical axis and horizontal axis as the casing. Adjustable casing spacers and rail guides are common in gravity carrier pipes, because they allow for a resetting of the line and grade independent of the casing.

When using casing spacers or centralizers, the bottom runners must be long enough to provide the minimum specified support to hold the carrier

Figure 2-19. For cutting and beveling pipe sections in the shop and or in the field, an oxyacetylene torch mounted in a "buggy" follows a track chain around the pipe; an alternative would be to use a plasma cutting machine.
Source: Courtesy of National Welding Corp.

pipe in place during grouting when the carrier pipe is subjected to buoyancy uplift forces. Typically, a minimum of 3 in. (75 mm) of annular thickness is needed to pump the backfill grout into the annular space. Some owners may require additional annular space. Casing spacers and centralizer legs must be designed to withstand high installation loads as the carrier pipe is pulled or pushed into position. It should be recognized that any spacer or centralizer with a single vertical leg at the 6 o'clock position (this occurs with an odd number of legs) may fail because the one vertical leg may take the full installation load. Figure 2-20 shows a typical configuration for a casing and carrier pipe. The invert legs should be at the 5 and 7 o'clock positions because the entire weight is shared by two legs. If a center invert leg is included (making an odd number of legs), then the entire weight will very likely rest on a single leg. This may result in an overdesign or overloading of the center leg, which leads to failure of the leg or failure of the carrier pipe if the wall has insufficient strength to take the effective point loading of the carrier pipe.

A specific consideration for gravity carrier pipes that use casing spacers to correct carrier pipe grade relative to the installed casing position is that carrier pipes may be subject to rotation during the installation process. This is particularly challenging for longer installations because it can be difficult to prevent this rotation, and the rotation could cause the carrier pipe to have an undesirable grade. If a gravity installation requires precise vertical

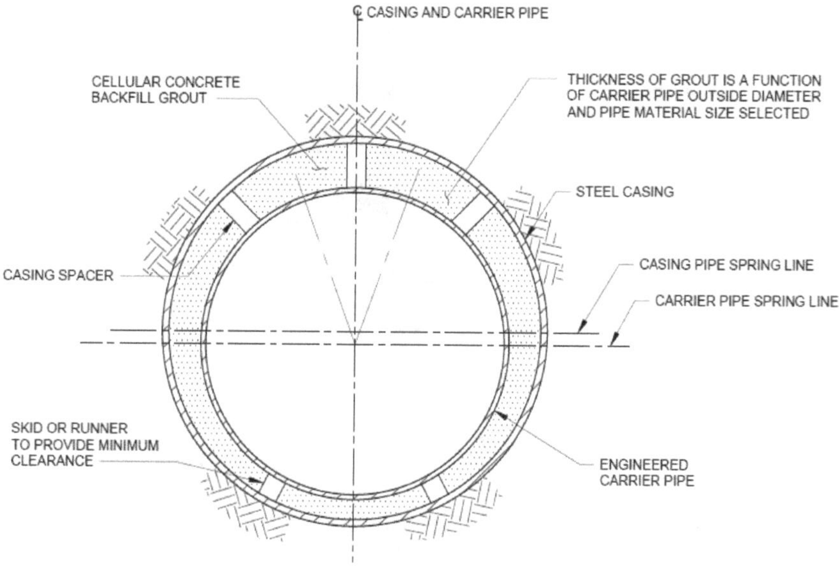

Figure 2-20. Typical configuration for a casing with the carrier pipe grouted in place.
Source: Courtesy of McMillen Jacobs Associates.

adjustments, it is easier use a preinstalled rail system set at the design grade followed by pulling or pushing the carrier pipe into the casing supported on the rails. This approach may necessitate ballasting the carrier pipe to maintain position during grouting, using lightweight cellular grout, or placing spring-type supports periodically located along the top of the carrier pipe to hold it in position.

Annular Space Backfill The annular space is the volume between the inside of the casing and outside of the carrier pipe for the length of the installation. The space is often backfilled with grout, control density fill (CDF), cellular grout, or blown sand. In some cases, the annular space is left open. Backfilling achieves the following:

- Stabilizes the carrier pipe position,
- Provides uniform bedding support to the carrier pipe,
- Increases the system capacity to support external loads,
- Prevents long-term buoyancy movement of the carrier pipe,
- Provides corrosion protection,
- Provides a separation between multiple carrier pipes,
- Prevents infiltration of groundwater and fines,
- Provides confinement to prevent casing or carrier pipe buckling or collapse,

- Provides a thermal sink to mitigate carrier pipe temperature changes,
- Provides thermal conductivity when required to remove heat from the carrier pipe, and
- Provides uniform support to prevent long-term sags and high points.

Backfill grout also prevents water from filling or flowing through the casing. If water is given an opportunity to flow into the casing, fines may be washed away from the overlying ground or, if sand is used as annular backfill material, the annular sand fill may be washed away, leading to casing failure or settlement at the surface. If the annular space is left open or filled with sand and there is contaminated groundwater, the annular space may serve as a conduit for contaminated groundwater movement.

Grout backfill should have the following attributes:

- Be capable of filling without developing voids,
- Have good flow characteristics to encapsulate the entire carrier pipe,
- Be viscous because it must be pumped,
- Maintain workability from time of mixing to injection,
- Be capable of maintaining aggregate (if present) in uniform suspension until set, and
- Be capable of displacing water without diluting the grout if water cannot be removed from the casing before grout placement.

These attributes are achieved by controlling the following backfill properties:

- Viscosity,
- Density,
- Suspension capacity,
- Gradation of aggregates (if aggregates are in the mix design),
- Set time,
- Temperature during curing (heat of hydration from pozzolan), and
- Electrical/thermal conductivity.

These properties can be measured and controlled by grout design and monitoring during the backfill grouting process to achieve the desired attributes. It should be noted that a pozzolanic (i.e., cementitious) grout backfill may generate enough heat during hydration to raise the grout temperature above 200°F, resulting in melting or deforming plastic pipe. To prevent damage to the carrier pipe, it should be filled with water (larger pipes) or filled with circulating cool water (smaller pipes) to absorb the heat from the hydration process. The high heat capacity of water provides an excellent heat sink and has been found to prevent significant temperature increase of the grout or carrier pipe during the curing process.

The fluid pressure during backfill pumping may lead to carrier pipe deformation or failure if the grouting pressure exceeds the unconfined critical buckling pressure of the carrier pipe. Injection pressure should be monitored, and an open vent should extend above the casing crown from the top of the high end of the casing to relieve the injection pressure.

For casings less than approximately 150 ft (46 m) and installed at a constant slope, grout may be pumped from the low end of a casing. For longer casings, grout placement should use tremie pipes with discharges located at periodic intervals along the casing. For casings with vertical curves, the tremie pipe discharge should start at the low point of the casing so grout will advance toward both ends and fully encapsulate the carrier pipe.

Annular grouting is not a routine construction process and should only be done by contractors experienced in annular grouting. Likewise, grout design is often tailored to the final use of the carrier pipe. For example, in the power transmission industry, the Electric Power Research Institute (EPRI) has a manual of practice on backfill grouting. Other industries may have developed special requirements and methods for backfilling, which the engineer should consult.

Grout pipes and associated joints must be pressure-rated materials and couplings to handle the grout pressures. Glued couplings are not normally acceptable and add significant risk to a project. Remember that the majority of construction cost has already been invested in the installation before grouting. Failure of the grouting will cause failure of the entire installation and could require a whole new installation.

Grout Mixing The grout mix source depends on the size of the project, the quantity of grout needed, the type of grout needed, and the grouting sequence. For larger projects, grout mixing can be performed by mixing plants set up on site, with dry mix grout and concentrated foam (cellular grout) being delivered in bulk. On moderately sized projects, grout mixing can be performed by self-contained, flatbed-mounted grout plants. On small projects, ready-mix trucks can deliver the premixed grout for direct pumping.

The grout mix design is typically considered to be a specialized mix and is not commonly available at a commercial mix plant. Annular space grout is designed for a specific purpose and one mix does not work for all conditions. A successful grouting program should be designed by a competent grout designer experienced in both grout mix design and in installation design and injection procedures.

Construction issues associated with annular space backfill grouting include the following:

- Bulkheads or lifts need to be constructed to contain the grout at the casing ends, and they must be capable of withstanding the hydrostatic pressure imposed by the fluid grout;

- Backfill grouting can be completed from the casing interior if it can accommodate personnel entry (typically 4 ft [1.2 m] minimum diameter);
- Access points to the casing will influence the sequencing and injection rate of the backfilling process;
- The potential for carrier pipe flotation will influence how the backfill grouting is staged and possibly the number of lifts required to fill the annular space; and
- The heat of hydration and the anticipated injection pressure in the annular space (not pump pressure) will dictate if plastic pipe needs to be filled with water before grouting.

Cellular grout is completed in lifts based on the capacity of the cell to survive and not collapse. Most cellular grouts collapse with 4 ft (1.2 m) of pressure. The maximum allowable lift height should be 80% of the pressure needed to collapse the grout.

If the casing is constructed with a slope, backfill grouting is performed from the downstream end of the system.

Backfill grout needs time to flow, which limits the amount of grout that can be injected in one day. Multiple lifts require planning. Some questions regarding design considerations for multiple lifts should be asked:

- Are the joints at the interface of different lifts vertical or sloped?
- Is a void space at the crown (i.e., bird's beak) above the bulkhead allowed when vertical joints are specified?
- Is one continuous lift for the top (i.e., crown fill) required?
- How will the grout sequence be staged to prevent flotation?
- If a plastic carrier pipe is used, how is the heat of hydration going to be controlled so that the pipe is not damaged or deformed?

2.4 PILOT TUBE GUIDANCE

2.4.1 Pilot Tube Installation

A pilot tube can be installed in advance of pipe ramming when the contractor needs to establish tighter control of line and grade for the pipe ramming installation. ASCE (2017) provides more details on the pilot tube installation process. Once the pilot tube is installed to line and grade, the contractor installs a series of step-up adaptors for the ramming process, like those shown in Figure 2-21. The step-ups are similar to those used with pilot tube guided auger boring. Multiple step-ups are needed because of the stiffness and strength differential between the final casing and the pilot tube. Without the step-ups, the trailing casing will likely shear from the pilot tube.

The following are primary advantages of using a guidance system:

- Allows for a straighter installation and a longer installation because of reduced frictional forces;
- Permits use of an overall smaller diameter casing because the installation is at the desired line and grade, which in turn reduces the volume of excavation and reduces steel and other material costs; and
- Reduces labor costs for checking of line and grade, welding, and clean out.

2.4.2 Step-Up Casings and Installation Sequence

Installation of step-up casings involves the following steps:

1. A small step-up is first attached to the pilot tube, and this assembly is pushed up to the shaft face. If there is no room in the shaft, the first step-up with casing may be rammed into the ground.
2. The ram is detached, and the next step-up with the larger casing is attached. The first and second step-ups are joined with a welded connection, so the contractor needs to plan for full circumferential access to complete the weld. It is also important that the connections are square and flush to minimize the introduction of alignment bias in the system during the ramming process.
3. After the second step-up is pushed up to the wall face, the last actions are to connect the casing to the final step-up, complete the burn-through of the shoring wall, and ram the entire assembly into the ground. The pilot tube will be received at the receiving shaft, so there should be good communication between the two shafts during the advancement and pilot tube retrieval phase.

Combining pipe ramming and the pilot tube method is a low-cost, fast, and effective way of installing casing in the ground on an accurate line and grade. Except for the ramming hammer and the pilot tube, the tooling is relatively inexpensive; in many cases the contractor adapts and reuses equipment, making it even more cost effective. For these reasons, more projects are using pipe ramming in conjunction with the pilot tube guidance.

2.5 TYPICAL APPLICATIONS

2.5.1 Road and Railroad Crossings

For many pipeline projects, construction involves open-cutting a trench, placing subbase foundation material, laying the pipe into the trench, and

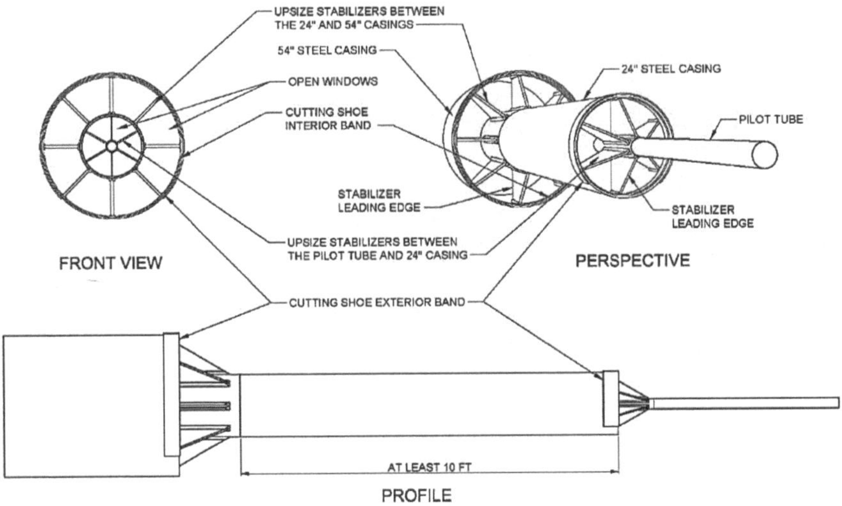

Figure 2-21. Step-up (upsize) adapters.
Source: Courtesy of McMillen Jacobs Associates.

then backfilling the trench. Often these linear projects will encounter physical obstacles that limit the use of open-cut trenching. These physical obstacles include but are not limited to

- Highways,
- Railroads,
- Utilities,
- Walkways and bike paths,
- Parks and golf courses,
- Waterways (environmentally sensitive areas, creeks, streams, rivers), and
- Airport runways and taxiways.

To cross these obstacles, a trenchless construction method is typically used. Pipe ramming can be economically and technically suitable for crossings from 100 to 300 ft (30.5–91.0 m) in length, with diameters up to 120 in. (3,050 mm). Other trenchless methods may be required when the crossings are longer, require a curved alignment, or are below the groundwater table (Table 1-1).

Because pipe ramming requires the use of a steel casing, does not require the ground to be excavated during casing advancement, and has a low potential for overexcavation, it is the preferred trenchless method of railroad companies and state DOTs. AREMA (2019) specifies pipe ramming as an approved construction method.

2.5.2 Culverts

Aging or undersized culverts can be replaced or repaired by one of two ways: rehabilitation of the existing culvert or new construction.

Rehabilitation does not remove the existing culvert material, but instead provides a new inner layer. ASCE (2009) outlines some of the in-place repair options available. The new lining material should be selected based on safety and life cycle considerations. Installation of any lining system will decrease the internal diameter, thereby potentially reducing hydraulic capacity. Sometimes the new smoother finish maintains or slightly increases the hydraulic capacity. However, a new inner lining is unlikely to meet the 100-year storm plus 1 ft (0.3 m) flood requirement.

Rehabilitation does not address or repair ovality, eroded bedding, lost invert, or backfill materials. The loss of material could be a contributing cause of a culvert's failure, which is evidenced by existing ovality.

Another problem with the rehabilitation approach is the construction sequence and timing. If the culvert is part of a natural flowing stream, damming upstream and bypass pumping will be required. This construction step is difficult to permit from an environmental perspective because aquatic life in the stream could be affected by the pumping, and the formation of deeper waters will change the ecosystem. In areas with well-defined dry and wet seasons, the culvert rehabilitation work could be scheduled for the dry season, but still could be affected by thunderstorms and sudden changes in weather.

In some instances, rehabilitation is just not possible (Syachrani et al. 2008). An existing culvert that is surrounded by unconsolidated or shifting bedding must be replaced. Lining will not address the effects of groundwater eroding the ground along the outside of the host pipe. Many of the problems with a culvert, like squatting or sagging, will not be corrected by inserting a new lining. A sagging culvert will still pond water, allowing the deposition of sediment and debris, which results in diminished hydraulic capacity. The result is the need for a full culvert replacement.

Culvert replacement may be accomplished by one of two approaches:

- Consumption (also known as pipe swallowing): The existing culvert is completely engulfed by a new larger diameter culvert and removed in its entirety; or
- Parallel construction: A new culvert is installed adjacent to the existing culvert and the existing culvert is then either rehabilitated, swallowed, or abandoned.

In the consumption/swallow approach, a new culvert (casing) is installed, completely engulfing the old culvert (Figure 2-22). The advantage is that the culvert's cross-sectional area is increased, increasing the culvert's flow capacity. The old culvert is entirely removed, and a portion of the bedding materials can be removed, providing a better soil-structure interaction for the new culvert. The size of the installation must consider how the space inside the casing will be adjusted to match the grade of the original culvert. Additional effective diameter may be required to accommodate the new grade and (for fish-friendly culverts) the aquatic-sensitive granular streambed material. In some conditions, only flow capacity is an issue. In these situations, an invert slab may be poured and different invert channel lining geometries cast to improve bedding material placement and stream throughput, and reduce scour potential. The new poured invert can be made flat, notched, or even roughened to allow the accumulation of some flow debris in the invert for environmental reasons and habitat development. Because the consumption/swallowing method removes the old culvert, any sagging in the original culvert st can be removed by increasing the diameter of the new casing.

If the culvert has flowing water, the water can be redirected into a new gravity bypass pipeline and routed through the existing culvert during construction, if there is room, or pumped around the construction. The consumption/swallowing method also allows the work to be completed from the downstream end. Depending on the amount of equipment used, the new culvert can be evacuated if a storm hits during construction. After the storm has passed, the construction can resume with minimal effort. In the consumption/swallowing approach, the work can be completed using pipe ramming, tunneling, or pipe jacking.

The consumption/swallowing method is useful when additional area to the left or right of the existing culvert is not available. Evaluation of the backfill material used around the original culvert is important to avoid hitting buried objects during construction. The consumption method will also reduce the amount of cover, so it should be evaluated.

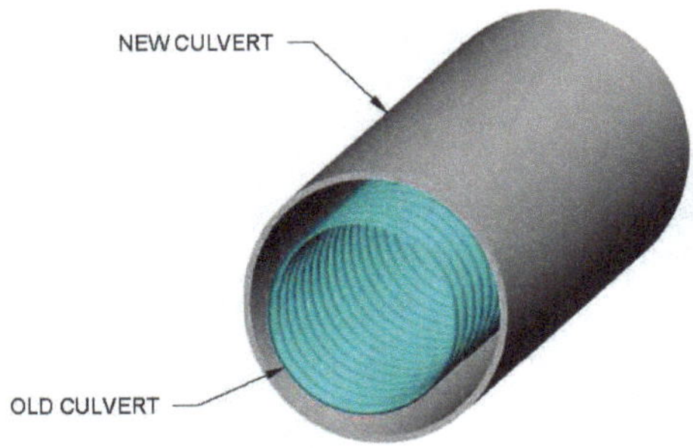

Figure 2-22. The original culvert is consumed in place by the new culvert, also known as pipe swallowing.
Source: Camp et al. (2010).

The second replacement method is the building of a new culvert parallel and adjacent to the old culvert (Figure 2-23). The advantage of this approach is that the culvert's effective cross-sectional area can be increased as needed to meet the flow capacity requirements. Alternatively, a smaller culvert may be installed followed by a slip-lined rehabilitation of the existing culvert, resulting in the same or larger flow capacity while reducing construction total cost. The invert of the new barrel can be set to the design profile, so a smaller effective overall diameter is needed with this approach. The old culvert remains in service during construction. Once the second barrel is in place, the old culvert can be replaced by the consumption/swallowing method, repaired, rehabilitated with a slip-line process, or abandoned if extra capacity is not needed.

If the stream has flowing water, the parallel approach allows the new barrel to be constructed while water flows through the original culvert. This is a huge benefit during construction. When the new culvert is completed, the water course can be redirected into the new barrel to allow consumption or rehabilitation work to occur in the old culvert.

Construction of a second culvert allows the work to be completed from whichever end has the best access. In the parallel culvert approach, the work can be completed using pipe ramming, tunneling, and pipe jacking.

2.5.3 Pipe Crushing

Pipe crushing is a method whereby a flexible corrugated metal pipe (CMP) culvert is consumed/swallowed (as described in Section 2.5.2) and replaced by pipe ramming a large-diameter casing over the existing pipe. Pipe swallowing and pipe crushing are cost-effective ways to replace smaller diameter pipes ranging from 24 to 60 in. (610 to 1,525 mm) in

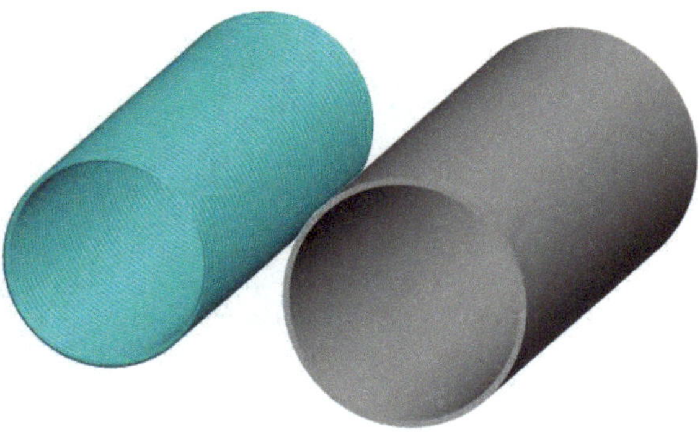

Figure 2-23. A new culvert is added on either side of the original culvert. Source: Camp et al. (2010).

diameter. However, extracting the CMP culvert with the spoils can be problematic. Attempts to push out the CMP can cause it to "accordion," further wedging it in place. Hooks or other fasteners simply tear away during attempts to pull it out.

The pipe (culvert) crushing variation of pipe swallowing takes advantage of the CMP's deformability. A "dummy pipe" containing long tapered blades begins the pipe swallowing process. As the dummy pipe and casing are rammed over the existing culvert, the blades inside the dummy pipe deform the CMP into a cloverleaf pattern (Figures 2-24 and 2-25). This process diminishes the existing culvert's circumference and decompresses spoils in the annulus between the old culvert and the newly installed, larger diameter casing, significantly reducing drag. The contractor can then use a backhoe, excavator, or similar piece of equipment to pull the crushed CMP and spoils clear from the newly installed casing.

Longer pipe swallowing runs and larger diameter culverts may still require extraction of the existing culvert in sections. However, use of the pipe crushing method usually makes it possible to remove much larger sections of CMP at a time, significantly reducing the total time and labor required to complete CMP extraction.

2.5.4 Slick Boring

An adaptation of pipe ramming that has primarily been used in the oil and gas market is called *slick boring*. This refers to the process of removing a sacrificial steel casing pipe by either pushing it out with the carrier pipe using the auger boring equipment or pulling it out.

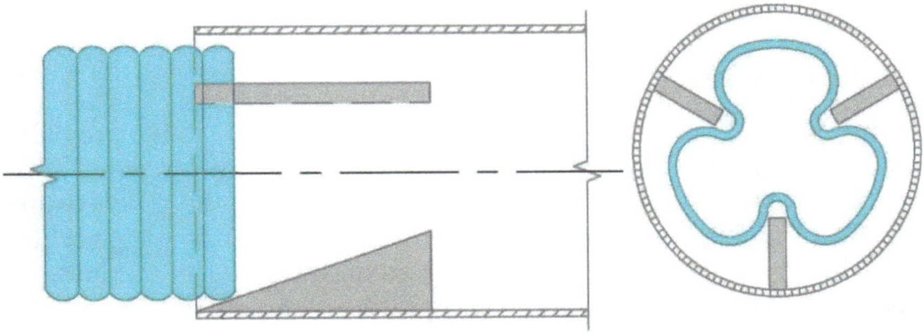

(a) (b)

Figure 2-24. Pipe crushing: (a) dummy pipe and tapering blades swallowing the old CMP culvert; and (b) associated cloverleaf pattern of the CMP inside of the new culvert after crushing.

Figure 2-25. Culvert crushing is a variation of the pipe swallowing technique especially designed for use on CMP culverts. The pipe is deformed into a cloverleaf-like shape during the ramming process for easier extraction. Source: Courtesy of Hammerhead Trenchless.

During the slick bore process the carrier pipe (usually a coated steel pipe used to convey oil or gas product) is welded to the back of the installed sacrificial casing. As the sacrificial casing is pushed (or pulled) out of the bore, the carrier pipe follows the established bore path and is pushed (or pulled) into place (Figure 2-26). This process is commonly used for relatively short crossings and allows a steel product pipe with its coatings and linings to be installed without being exposed to the stresses and damages associated with ramming and the auger clean-out process. The sacrificial casing is usually removed by either pushing/ramming the product pipe in after the casing (similar to pipe extraction) or by pulling the casing out in front of the product pipe (as shown in Figure 2-26).

2.5.5 Pipe Extraction

Pipe extraction is a newer method that has been used to replace older steel pipelines with new steel pipelines using pipe ramming. In pipe extraction, a section of new pipe is attached/welded to one end of an existing pipe (Figure 2-27). A hammer is attached to the end of the new pipe, which is used to push the old pipe out of the ground, or the hammer may be attached to the old pipe and the new pipe pulled into the existing hole by the old pipe as it is removed. If the pushing-in process is used, then a static pull assembly may be attached to the other end of the old pipe/casing to assist in pulling and to provide static tension to limit the bounce-back effect of the hammer.

As new sections of the steel pipe are rammed into place, the old pipe is cut off and the static pull assembly is reset. Additional sections of new pipe

Figure 2-26. Slick boring concept.
Source: Courtesy of TT Technologies.

Figure 2-27. Example of a pipe extraction in which a portion of the existing pipeline is exposed, and welded to the new steel pipeline. Using the pipe rammer, the existing pipe is then extracted from the ground and replaced by the new pipeline.
Source: Courtesy of Hammerhead Trenchless.

are welded to the previous new section of pipe, the hammer is reset, and the pipe extraction process continues.

2.5.6 Combining Pipe Ramming and Horizontal Directional Drilling

Some of the most interesting and impressive trenchless projects have been accomplished using HDD. However, drill operators and manufacturers are finding new and creative ways of addressing particularly difficult projects and situations by using pipe ramming to assist with the directional drills. The techniques are changing the way drillers approach projects and respond to problems on the job.

Several techniques have been successfully used to help prevent pipe pull failures in directional drilling projects or salvage bores by using pipe ramming. Properly configured pipe rammers can be used to salvage pipes, remove stuck drill stems, assist drills during pipe pullback, or address other problems that may hamper the pipe installation. Drilling contractors throughout North America have successfully used the directional drilling assist techniques described next.

Bore (Product Pipe) Salvage Pipe ramming may be used to salvage a stuck carrier pipe. For this procedure, a hammer is attached to the exposed end of the partially installed carrier pipe in an orientation that tends to pull the pipe from the ground. This can be accomplished by using an extraction cradle (Figure 2-28) or a fabricated sleeve (Figure 2-29). A winch or other type of pulling device is used to assist the hammer during the pipe removal operation. The percussive power of the hammer is often enough to free the stuck carrier pipe and to then allow it to be readily removed from the ground.

Drill Stem Recovery There are two possible configurations for applying a hammer to assist in drill stem recovery. Depending on the situation, contractors can directly pull the drill stem from the ground using the power

Figure 2-28. Extraction cradle tool.
Source: Courtesy of Hammerhead Trenchless.

Figure 2-29. Bore salvage.

of a ramming tool or, if the stem is still attached to the drill rig, they can use the ramming tool power to push on the opposite end of the stem to assist with the drill rig pullback action (Figure 2-30).

Pullback Assist The pullback assist technique helps install the carrier pipe in problematic situations. For example, when drilling underwater, in loose flowing ground conditions, or when there is loss of drilling fluid circulation, a condition known as hydrolock can occur. Hydrolock results when the pressure at the leading end of the carrier pipe restricts its forward movement. Alternatively, earth pressure along the side of the pipe caused by partial collapse of the bored hole may cause additional frictional drag, thus inhibiting the pipe movement. In such cases, the required pull forces may exceed the drill rig's pullback capability or the carrier pipe's tensile strength. The percussive pushing action of a pipe rammer applied at the tail end of the carrier pipe may be used to help free the immobilized pipe.

The pullback assist technique has been successfully used on steel pipe as well as HDPE. The technique can be used initially as a precaution in anticipation of possible problems, such as those described previously, or after the pipe has become immobilized (Figure 2-31). Response time, however, is a key factor. The rate of success greatly improves as the response time decreases. Therefore, many drilling contractors bring ramming equipment to the directional drilling sites so they can respond quickly to problems that may develop.

Conductor/Washover Casing The success of a drilling operation can often be determined before it is initiated. If the ground conditions at the planned entry or exit point are problematic, the success of the entire project may be jeopardized. In such cases, a conductor casing may be installed at entry and/or exit locations in advance of HDD work to extend a ground support casing through the problematic ground conditions.

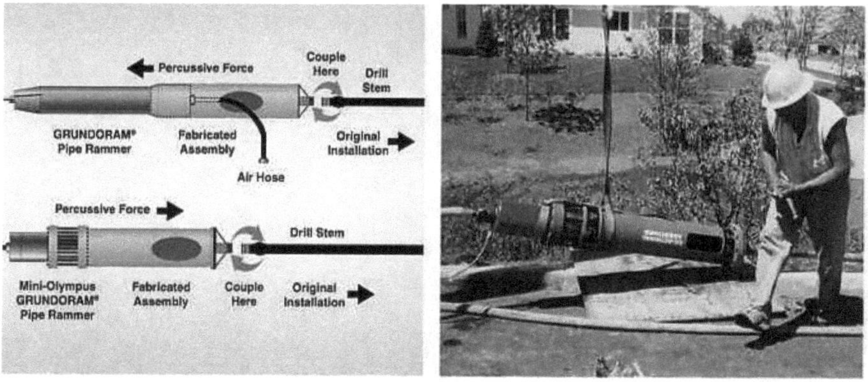

Figure 2-30. Drill stem recovery.

Figure 2-31. The pullback assist technique.

The technique differs slightly from the preceding methods because it is incorporated into the initial boring plan, rather than being deployed only in the event of a problem that may arise at some stage. In this method, a clear casing-supported pathway is provided through the poor ground conditions. Areas with loose, unsupported ground and ground with cobbles and boulders are prime candidates for this method.

During the conductor casing installation, open-face steel casing is rammed into the ground at a predetermined angle until desirable ground conditions are reached. The casing size is typically sufficient to permit the largest anticipated drilling tool to pass through the casing. The spoils are removed from the casing with a drill-post type auger or core barrel. A centering casing may then be placed inside the conductor casing to center the pilot drilling in the conductor casing. This centering casing should extend a distance from the end of the conductor casing to prevent ground washout at the end of the conductor casing when tools are retrieved. Drilling then proceeds within the centering casing, beginning at a point where more desirable ground conditions are encountered.

In addition to assisting drilling operations at the start and end of the installation by supporting the ground, the conductor casing provides a barrier to prevent drill fluids from leaking to the ground surface through the poor ground and shallow cover situations. The conductor casing may also serve as a low friction section, facilitating pullback (Figures 2-32 and 2-33).

Figure 2-32. Installation of a conductor casing.
Source: Courtesy of TT Technologies.

Figure 2-33. Conductor/washover casing installation.
Source: Courtesy of Hammerhead Trenchless.

REFERENCES

AREMA (American Railway Engineering and Maintenance-of-Way Association). 2019. "Guidelines for railroad grade separation projects." Lanham, MD: AREMA. http://www.arema.org.

ASCE. 2009. *Trenchless renewal of culverts and storm sewers*. MOP 120. Reston, VA: ASCE.

ASCE. 2017. *Pilot tube and other guided boring methods*. MOP 133. Reston, VA: ASCE.

AWS (American Welding Society). 2014. *Specification for welding procedure and performance qualification*. ANSI/AWS B2.1. Miami: AWS.

AWS. 2020. *Structural welding code—Steel*. ANSI/AWS D.1.1. Miami: AWS.

Camp, C., G. Boyce, and A. Tenbusch. 2010. "Culvert replacement using pipe ramming, tunneling, or pipe jacking." In *Proc., North American No-Dig 2010*, Paper A-5-05. Chicago: North American Society for Trenchless Technology (NASTT).

EPRI (Electric Power Research Institute). https://www.epri.com/#/search/backfill%20grouting/?to=1572120825365&from=607455974634)

Meskele, T., and A. Stuedlein. 2015. "Drivability analyses for pipe ramming installations." *J. Geotech. Geoenviron. Eng.* 141 (3): 04014107.

Syachrani, S., V. Rai, and H. S. Jeong. 2008. "Evaluation of DOT practices in the United States for rehabilitation of transportation drainage structures." In *Proc., North American No-Dig 2008*, Paper F-1-01. Dallas: North American Society for Trenchless Technology (NASTT).

Werner, S. 2003. "Installation of culverts by pipe ramming." In *Proc., Ann. AREMA Conference*, AREMA, Chicago, October.

CHAPTER 3
PLANNING

3.1 INITIAL CRITERIA

Pipe ramming is a specialized construction method used typically for undercrossing of railroads and highways to install utilities, transmission pipelines, and culverts. It is suitable for conditions in which open-cut construction methods are not cost effective, are too disruptive, or are not technically possible. As outlined in Chapters 1 and 2, pipe ramming is typically used to install pipelines above the groundwater table that are between 8 and 144 in. (205 mm and 3,660 mm) in diameter for distances of approximately 200 to 300 ft (61 to 91 m). Installations below the groundwater may be possible and are limited by ground type, depth below the groundwater table, ground improvement methods, or specific equipment used.

When construction occurs in rock conditions, other trenchless methods may be better suited to install the pipeline or casing, such as

- Horizontal directional drilling,
- Horizontal auger boring with a rock head,
- Microtunneling, and
- Utility tunneling.

There are significant differences between these construction methods, both in application and in results. The engineer should review each of these methods to determine whether it can meet the project requirements and be used successfully under the specific conditions that are anticipated during the pipe installation process.

The following parameters will help to determine the applicability and constructability of a potential pipe ramming project:

- Geotechnical conditions and geology, including groundwater;
- Potential for encountering objects (both natural and remnants of past construction) along the proposed alignment;
- Environmental conditions, including potential for contaminated ground or groundwater;
- Available work space for shafts or portals;
- Required drive lengths;
- Casing or carrier pipe diameters;
- Cost and construction schedule;
- Alignment;
- Site access;
- Depth;
- Grade;
- Installation tolerances;
- Effects on surface activities;
- Location of existing, abandoned, or proposed utilities;
- Rights-of-way; and
- Location of existing, abandoned, or proposed structures.

The engineer of a project is encouraged to consider all trenchless methods before concluding that the pipe ramming method is the best construction method available. MOPs have been written by ASCE and others for various construction methods. Useful references are provided at the end of Chapter 1.

If the engineer responsible for the project does not have a strong background in trenchless design, an engineering firm that specializes in trenchless designs should be consulted to provide a peer review early in the planning/design process to help make good design choices.

3.1.1 Casing Diameter and Grade

The casing diameter must be of sufficient size to meet the project requirements. For gravity sewers, the grade should be established to meet the project requirements and to accommodate the carrier pipe diameter selected. Flow capacity (Q) and cleansing velocities must be considered for gravity sewers. For pipe ramming, the selection of the casing to be installed is then made considering the design tolerances of the line and grade required of the carrier pipe.

Design grades and tolerances are typically specified for gravity systems. Grades and tolerances are less critical for pressure pipelines and dry utilities and conduits. When grades are relatively flat or critical, or when unfavorable ground conditions exist, a pilot tube-guided pipe ram, a two-pass approach, or a large oversized casing should be considered. The two-pass approach first installs a larger diameter casing and then installs the carrier pipe inside the casing at the required line and grade. The size of the casing must include consideration for alignment and grade installation tolerances and space for backfill grouting.

For linear pipeline projects, the engineer should consider including an additional drop of 1.2 in. (30 mm) or more, plus the grade tolerance at each manhole location, to assist in maintaining the design grade during construction. Without this additional drop across the manholes, installation of the carrier pipe within the casing must be more accurate. Over the length of a project, these drops can be used to help maintain the overall grade in the pipeline if one section becomes misaligned during construction.

3.1.2 Depth

The following items should be considered when determining the installation depth:

- Gravity flow or pressure pipeline;
- Required start and end points as well as intermediate pick-up points or tie-in points;
- Project performance (i.e., hydraulics, connections, operations, and maintenance);
- Geotechnical conditions, geology, and groundwater level;
- Existing buried infrastructure;
- Future construction and changes to the ground surface elevation;
- Potential and tolerance for vibration-induced settlement or heave; and
- Potential and tolerance for lubricant reaching the surface.

Increased depth does not affect the pipe ramming method as it does open-trench construction. The major costs associated with deeper pipe rammed installations are shafts. Therefore, when possible, the installation elevation for the pipe ram can be adjusted to take advantage of better ground conditions to avoid other infrastructure or to meet future needs. Adjusting depth is not always possible for gravity systems with flat slopes, fixed end points, high groundwater table, or when tying into an existing system.

Selection of minimum ground cover should take into consideration the ground conditions and use of the surface space above the casing. Minimum ground cover is typically a function of gravity versus pressure flow, casing/carrier pipe diameter, geology, and traffic loading. As the ground cover decreases, the risk of surface settlement and heave increases. Ground treatment may be required to control surface settlement or heave with minimal ground cover. Because the pipe ramming process can occur with no active excavation during the casing installation, the ground cover for a typical crossing may be only a few feet (or a meter).

Typically, a minimum cover is needed to allow sufficient depth for the ground to arch over the new casing installation. The minimum depth of cover is determined based on casing outside diameter, existing ground conditions, and other site conditions. Evaluation and calculation of potential vibration-induced settlement, heave, and lubrication pressure should be performed by the engineer to assist in the selection of the appropriate ground cover. Potential ground settlement or heave and the owner's tolerance to the attendant risk should be taken into consideration when establishing the depth and the amount of ground cover. These considerations are addressed in more detail during the design process (see Chapter 5).

3.1.3 Routing

Routing of the new pipeline installation can be affected by many factors, including

- Required start and end points as well as intermediate pick-up points or tie-in points;
- Geotechnical conditions, geology, and groundwater;
- Surface topography;
- Existing utilities;
- Physical buried objects either placed naturally or the result of construction;
- Available rights-of-way;
- Maintenance of traffic during construction;
- Surface access; and
- Contaminated ground and groundwater.

Refer to *Pipeline Route Selection for Rural and Cross-Country Pipelines*, MOP 46 (ASCE 1998) for more information.

Pipe ramming has been used as a cost-effective alternative to, or in conjunction with, open-cut construction and other trenchless methods where crossings are required, surface disturbances are to be limited, trench support is challenging, conflicting utilities are encountered, or deep trenches are required.

Areas of contaminated ground or groundwater should be avoided if possible, especially at shaft locations. If they cannot be avoided, a trenchless installation is often a good option because it minimizes the volume of material being excavated compared with open-cut trenching. Special handling and disposal of contaminated spoils should be considered.

In urban areas, pipe ramming portals, shafts, and working spaces are typically located perpendicular to the crossing. Additional working areas may be required for casing and carrier pipe storage, other equipment, and office trailers.

Underground utility density is also an important consideration in determining the route. Refer to *Standard Guidelines for the Collection and Depiction of Existing Subsurface Utility Data* (ASCE 2002) for guidance on the collection and presentation of existing utility data. The preferred trenchless route and alignment should minimize effects on existing utilities in terms of shaft placement, emergency access routes, and ground movement caused by surface settlement or heave. Trenchless methods in general provide lower impact solutions and options not typically available for open-cut trench construction.

3.1.4 Portals, Shafts, and Working Space

Pipe ramming operations typically involve crossing an embankment where the entry and exit points are at or near the ground surface (i.e., portals). If the pipe ramming operations are deeper, then the pipe ram proceeds from a launching shaft to a receiving shaft. The launching side of the crossing needs to be sized to accommodate the full casing length with the hammer attached and the equipment needed to clean out the casing. The size of the launching working space will be a function of the casing length plus the size and length of the hammer plus room for joining the casing segments together.

During the planning stage, working space on the surface to accommodate support equipment adjacent to the launching shafts should be identified. Space is required for

- Shaft and portal construction, including staging and handling of shoring materials and excavated spoils,
- Power source,
- Air compressor(s),
- Lifting equipment,
- Mixing tanks for lubrication and grouts,
- Casing and carrier pipe storage and handling,
- Temporary spoils handling and storage, and
- Other equipment necessary to support the pipe ramming operations.

Launching shafts or portals should be of sufficient size to accommodate the earth support system, the hammer required for the casing diameter being installed, the length of casing, shaft access devices for personnel, safety equipment, tools, and personnel work space (Figure 3-1).

Receiving shafts or portals need space for removing spoils and construction of the final structure. Unlike other trenchless methods, no excavation equipment is retrieved from the receiving shaft. If a pilot tube is used for

Figure 3-1. Typical pipe ramming site setup with guide rails, casing, attachment, hammer, hoses, and lubrication tank. Compressor is located on the surface adjacent to the portal.
Source: Courtesy of Michelle Macauley.

casing guidance, space will be required to retrieve it and perform any step-ups.

To the extent possible, shaft and portal locations should be selected that avoid overhead and subsurface utilities. For all shafts, maintenance of traffic during construction (including access for emergency vehicles) should be considered.

Shafts or portals are typically located at changes in alignment (both horizontal and vertical), changes in carrier pipe diameter, changes in geotechnical conditions and geology, or topographic low spots. Shaft spacing and locations should also be based on practical drive lengths (Table 1-1) and project design constraints. The location of shafts may be dictated by the required distance between maintenance access points/manholes to work with the owner's cleaning and inspection equipment.

Shafts constructed far in advance of the casing or carrier pipe installation can be covered with decking, if necessary, to address public safety or traffic concerns. Excavation of the launching and receiving shafts in advance of pipeline construction is useful in confirming geotechnical subsurface conditions before installing the casing. Early excavation of shafts should be considered when complex or difficult ground conditions are expected.

The shaft construction usually affects the public longer than the actual casing and carrier pipe installation operations and can present risks to the owner, such as settlement, break-in/break-out failures, shaft wall-to-floor failures, and utility disruptions. When locating shafts consider the utilities (overhead and underground), settlement of sensitive features, traffic disruptions, vibrations during shaft and portal construction, and disruptions to residences and businesses.

A launching shaft is in constant use during casing and carrier pipe installation and contains crews and equipment that must be protected. The launching shaft should be protected from flooding and cannot normally be covered, other than with a removable security cover during nonworking hours. The dimensions of the launching shaft need to take into account walers, cross bracing, thrust blocks if needed, entry/exit seals, ladders, ventilation ducts, sump pumps, the length of the hammer and hose connections, the casing segment length, and the equipment necessary to clean out the installed casing. Vertical space above and adjacent to the shaft or portal should be free of overhead utilities. Most shafts and portals will require crane access to build them and to position and remove the hammer, casing, carrier pipe materials, and support equipment. Analyses of safety hazards should be completed for all critical lifts. Overhead wires should be identified before mobilization of equipment. Attention should be paid to the latest Occupational Safety and Health Administration (OSHA) regulations regarding maintaining safe distance between cranes and lifting equipment and overhead power lines.

The size of the staging area will vary between shaft and portal construction and casing installation. During shaft and portal construction, the working space is larger because access is required around the entire shaft or portal. If the working space is located in a roadway, additional lane closures may be required. For planning purposes, the engineer should allow for staging areas consisting of 2,000 ft^2 (190 m^2) for launching shafts and portals and 500 ft^2 (43 m^2) for receiving shafts and portals. The launching shaft and portal are typically longer and narrower. The length of the launching shaft or portal is a factor of the casing and hammer lengths selected and the associated setup.

During casing installation, the launching shaft or portal staging area needs space for cranes, loaders, the air compressor, spoil removal equipment, utility pipes, casings, and trucks. The staging area adjacent to the receiving shaft needs space for cranes, spoil removal equipment, and trucks. These staging areas will be smaller than that needed for shaft or portal construction. Staging areas may be further reduced in size by storing casing and carrier pipe in other secondary work areas and using gantry cranes. Effects of work zones on traffic and other surface activities should be considered.

In planning the geotechnical investigation related to shaft or portal construction, consider that shaft or portal support elements often extend below the actual casing elevation. Geotechnical borings should extend below the shaft or portal support embedment and casing depth. When complex geology is present, consider locating shafts and portals at the interface between disparate geology to allow for modifications or changes to trenchless construction methods to better suit the different geology.

Preliminary drive lengths should be based on calculations performed by a professional engineer. Calculations should consider the project geology, casing diameter, casing thickness, accuracy requirements, cutting shoe design and survivability, and known buried objects. Engineers can also consult with other trenchless engineers, equipment manufacturers, and contractors, and review published case histories to establish drive lengths for the casing diameter being considered.

3.1.5 Sustainability

Sustainability comprises a set of environmental, economic, and social conditions in which society as a whole strives to maintain and improve the quality of life indefinitely, without degrading the quantity, quality, or availability of natural resources and ecosystems.

The pipe ramming method minimizes environmental impacts (such as dust), minimizes the volume of excavated materials, and reduces traffic effects (such as traffic delays and detours). The pipe ramming method provides greater sustainability benefits for owners and society than open-trench construction (ASCE 2019).

It is ASCE's policy to promote the goal of sustainable civil infrastructure. As stewards of the built environment, civil engineers lead the way in planning, designing, and constructing sustainable infrastructure. The pipe ramming method offers a sustainable alternative to open-trench construction by avoiding disturbances of ground surfaces, which would require reworking of roads, railroads, landscapes, or buildings. When possible, the spoils from the pipe ramming operation should be used directly (or indirectly) as backfill.

The carbon footprint of a trenchless project is typically less than that of open-cut construction methods and can be calculated using a carbon footprint calculator. Carbon footprint calculators for construction are available on the internet. The volume of excavated material for pipe ramming is much less than for open-trench construction of the same pipeline reach. This reduced excavated volume results in fewer truck trips for pipe ramming, and this makes trenchless methods a sustainable alternative.

3.1.6 Regulations

The engineer should identify and meet federal, state, and local regulations when planning and designing the project. At a minimum, the engineer needs to identify the following:

- Need for environmental documentation and permits,
- Local regulations for discharges and handling of excavated materials and groundwater,
- Erosion control measures,
- Noise limitations, and
- Work hour limitations.

3.2 SITE CONDITIONS

The owner and the engineer should visit the proposed site during the initial planning phase and periodically throughout design to identify site conditions and constraints. A survey of existing site conditions, both surface and subsurface, should be performed in the early stages of planning. This information is useful in determining the feasibility, alignment, grade, and physical constraints of the proposed pipeline project. This planning work is independent of the construction method selected. Individuals who understand constructability issues should be consulted and included in the site visits.

3.2.1 Land Use

To provide insight into anticipated subsurface conditions, potential future conflicts, and other potential problems, an investigation should be undertaken into the history of the area proposed for the pipe ramming operation as well as current zoning and plans for future development. A review of historical drawings, construction records, land-use maps, and aerial photographs can provide insight into past land use that might affect the feasibility and related risk of using pipe ramming. Review of aerial photographs, taken in different eras, can be particularly useful in assessing changes in land use and investigating evidence of previous construction. A survey of the project corridor must be done to identify existing infrastructure above and below ground level. Existing property access and traffic conflicts along the corridor need to be identified, reviewed, and understood. To the extent possible, potential future land uses that could restrict the location of structures or affect the alignment need to be identified. Additional site visits may be required in areas of rapid development.

3.2.2 Easements and Rights-of-Way

The availability of temporary or permanent easements, rights-of-way, and, if necessary, property acquisitions should be determined during the planning process. Early determination of unacceptable corridors may require changes in the alignment. Alignment changes are typically less costly and less disruptive to the project schedule when done early in the planning process. For some pipeline projects, the use of pipe ramming may be the only acceptable method for construction within an easement.

The project owner and the engineer should evaluate different alignments. Consider keeping the alignment alternatives within the public rights-of-way rather than extending onto private property, which would require acquisition or easement. The cost and schedule of acquisition of a private property or easement must be considered. A comparison of property/easement cost versus potentially longer and more challenging alignments that stay within public rights-of-way should also consider long-term maintenance cost. Efforts to acquire the needed private rights-of-way or easements should start as early as feasible to avoid delaying the project. Acquisition of easements and rights-of-way are often critical to the construction schedule.

Permanent easements for shafts and surface structures should take into account future maintenance and repair needs. Temporary construction easements are often required for access to the site, support of equipment, and storage of materials during construction. A drawing of the site (with all support materials, equipment layout, and shaft locations) should be developed to aid in determining the construction footprint and property needs.

Depth of the pipeline and geotechnical conditions should be considered when evaluating future impacts and property easement costs. When reviewing and acquiring easements, consider width and lengths for future maintenance or repair needs and the most likely inspection and repair methods that will be used. The diameter of the carrier pipe installed in the rammed casing often dictates the trenchless repair method, which in turn dictates future needs for access and permanent easement.

In some jurisdictions, a subsurface easement can be obtained rather than acquiring the property or obtaining a surface easement. Obtaining subsurface easements is sometimes easier and less costly than surface easements, but they come with restrictions. When a subsurface easement is obtained, a temporary construction easement may also be required for access to the pipe ram envelope from the surface. Subsurface easements could also affect future maintenance and operations.

3.2.3 Underground Utilities and Other Facilities

Underground utilities and other facilities along or crossing the pipe ram alignment must be identified during the planning process in accordance with state utility-protection laws. Use of ASCE 38 (ASCE 2002) is recommended, along with the Common Ground Alliance's *Best Practices* guide (2019). Utilities and other facilities within the proposed alignment can be affected in a number of ways depending on geology, adjacent underground structures, and material of construction of the existing utility. During the planning phase, it is important to identify those utilities that will have a significant cost, time, or fatal flaw impact on the trenchless component of the project. Professional judgment is required to determine how much utility investigation is necessary, but identifying known and suspected utility owners within the pipeline easement and asking them for their records is an important first step. At a minimum, according to ASCE 38, a Quality Level D (QLD) (records research) utility map should be produced. If an accurate topographic plan has been generated at this stage of the project, this QLD utility map can be upgraded to a Quality Level C (QLC) (records research correlated to an accurate site survey) version at minimal cost and time. If high-risk utilities are identified through this process, further actions may be prudent—such as upgrading specific utilities to Quality Level B (QLB) (surface geophysical methods) and Quality Level A (QLA) (exposure and test holes).

All states have a one-call planning or design ticket. In some cases, using these services may supplement or replace the use of ASCE 38, which may not identify all utilities or provide the detail necessary for the exercise of prudent professional judgment. The engineer must also consider that other objects (such as abandoned utilities, concrete thrust blocks, concrete encasement, manholes, valves, foundations, and vaults) may affect the project alignment. Trying to install a pipe ram through previous excavations or backfill may have an adverse effect on the grade of the casing, even if there is no direct conflict with an existing utility. For example, backfill material or bedding from an existing trench could create a mixed-face condition in an otherwise homogeneous material.

The use of the QLD map, the QLC map, and/or the one-call ticket allows the engineer to make planning decisions on alignment, depth, and necessary utility relocations prior to pipe ramming, as well as other risk-management issues.

In some cases, a more complete subsurface investigation may be warranted. The combined use of geophysics and test holes may be prudent. ASCE 38 contains information on the current capabilities of advanced geophysics to image utilities and other features. However, this step in utility identification is usually reserved for the design stage of the project. Particular care should be taken, and additional locating effort should be made when crossing utilities installed by the HDD method because of the variable installation depths inherent with the method. Older HDD installations have greater variability and uncertainty of marked locations.

3.2.4 Buried Objects

The potential for encountering buried objects along or crossing the pipe ram should be identified and evaluated during the planning process. The pipeline alignment should be adjusted to avoid known buried objects. In the planning process, buried objects should be defined, and the methods for resolution should be addressed. Additional assessments of potential buried objects within the planned pipe ram horizon should be undertaken during the geotechnical investigation.

Buried objects may include remnants from past construction or be natural in origin. Examples of remnants from past construction are building tiebacks; debris infill; abandoned utilities; wood, concrete, and steel piles; old foundations; shoring from past excavations; utility trenches; and past excavations. Examples of naturally occurring objects are trees; tree trunks; root systems; cobbles; boulders; voids filled with material that differs, materially and substantially, from the surrounding ground conditions; and flood debris.

A buried object encountered during the work that stops or impedes the installation of the pipe ram along its design path within permissible tolerances is an obstruction. If a pipe ram encounters an obstruction, depending on the ground conditions and casing diameter, the spoils in the casing can be removed and an inspection made at the leading edge of the casing. If a new alignment is selected, typically the casing is moved horizontally to the left or right of the first installation. Usually, the vertical position and grade remain unchanged. A shaft or other access from over the alignment could be used to remove the buried object to allow forward progress of the pipe ram along the original alignment.

Buried objects have been known to stop other trenchless methods because the equipment has a closed face with no to limited access. Pipe ramming with an open-ended casing has been used successfully in consuming a stopped trenchless installation and associated buried object(s) (Stuedlein and Meskele 2014).

3.2.5 Contaminated Ground

The type, location, and degree of contamination and the extent of contaminated ground should be determined to assess the effect of the contamination on the final alignment and construction materials, including carrier pipe and joint materials. Contaminated ground should be avoided, if possible. Because pipe ramming is typically done above the groundwater table, encountering contaminated groundwater is less of an issue. Disposal of contaminated ground and groundwater encountered during excavation will increase the overall cost of the project. However, the volume of contaminated material excavated, handled, and disposed of during trenchless installation is less than during open-cut construction. Trenchless installation therefore provides an economic advantage over open-cut methods. This advantage could result in an overall lower project cost.

Realignment of the project should be considered to avoid the contaminated ground and to avoid affecting any associated contaminated plume.

Explosive gases and volatile contaminants should be taken into consideration when determining whether pipe ramming is appropriate, selecting the method of shaft construction, and selecting the type of ventilation in the shaft and within the pipe rammed casing.

Consultation with specialty environmental and regulatory experts should be considered. Some jurisdictions (national, state, and local) require the preparation of an environmental baseline report (EBR) to document the location of contaminated materials, their potential impact on the environment, and required mitigation methods. The EBR should be considered a contract document, if used.

3.2.6 Special Considerations

Special project considerations should be identified early in the planning process. Special considerations may include the following and are discussed subsequently:

- Limited work hours;
- Restricted access;
- Environmental restrictions;
- Noise limitations;
- Dust control;
- Light limitations;
- Odor control;
- Traffic disruptions;
- Haul routes and disposal sites;
- Service interruptions;
- Business disruptions;
- Visual impacts;
- Potential settlement of existing facilities;
- Utilities underground and overhead;
- Disposal of lubricant, spoils, and groundwater;
- Bypass pumping; and
- Public events and construction moratoriums.

Limited Work Hours Projects located in residential areas may be subject to work hour restrictions by local ordinance. Work hours are typically restricted to daytime hours when work is being done in residential areas.

Restricted Access Construction activities for pipe ramming are focused around the shaft locations. Shafts should be positioned in areas offering access for trucks, employees, and supplies. The shaft sites should be fenced to restrict entry by the general public. Anyone entering a construction site should be required to check in with the contractor and wear PPE while on site.

Environmental Restrictions The pipe ram can be located near environmentally sensitive areas. Shafts and associated work sites should be located in such a way as to avoid the defined limits of the environmental area. Some environmental areas may have nesting birds or other wildlife that should be considered and protected. If an environmental document has been prepared for the project, the document may contain mitigation requirements to be followed. Restrictions may limit the time of year when the work can occur.

Noise Limitations Noise will be generated from the pipe ramming process. This noise will include the hammering of the steel casing into the ground for the casing length, the hum from generators, and backup alarms on trucks. The noise generated from the hammering will stop after each casing length is installed. Depending on the casing joint system used, the hammering of the casing will occur two or three times a shift. Additional cost may be needed to help mitigate or muffle the noise generated.

Dust Control Pipe ramming involves excavation of the ground. Depending on the moisture content of the spoils, the excavated material can become dry and cause dust to be formed. If the spoils are too wet, mud and fluid spoils can be deposited on the ground surface, where they will eventually dry out and also form dust. Trucks entering and exiting the sites can transport the dust to surrounding areas. Dust control may include sweeping the road surfaces on a regular basis, using water trucks to dampen the dust, placing covers over the trucks as they exit, and washing the truck tires as they leave the site.

Light Limitations The pipe ramming method involves the forward advance of the casing into the ground. Sometimes a 24 h operation is needed to keep the casing moving forward. In these circumstances, lights will be needed at the portal/shaft locations to allow work to continue through the night. The setup and orientation of the lights should provide sufficient illumination for the workers without disturbing adjacent residents. Screening may be required as a mitigation for light and visual effects.

Odor Control Odors may be noticeable to individuals passing by the work sites. Odors may be natural from organic clays, from sewage, contaminated ground, or construction equipment. Mitigation methods to reduce the effects of odor to the public should be considered based on the source.

Traffic Disruptions The pipe ram and the associated portals or shafts are often located within public rights-of way or in city streets. Because the equipment is modular, it can be set up in a variety of configurations. Occupation of the traffic lane requires redirection of traffic past the portals/shafts and the work sites. Management of traffic can disrupt traffic flow. Traffic control plans should be developed for all projects located in city streets. The engineer of the project should look for portal and shaft locations that minimize traffic disruptions or consider traffic detours when appropriate.

Haul Routes and Disposal Sites Spoils from the excavation process will be generated and must be disposed of. Owners may have to identify potential disposal sites for the contractor, especially if contaminated or hazardous material is likely to be encountered. Once a disposal site is identified, haul routes should be identified. Additional truck will be generated on a daily basis as trucks enter and leave the shaft sites.

Service Interruptions Pipe ramming is typically used to install casing under a highway or railroad. The process of adding service connections to the new pipeline will be completed elsewhere.

Business Disruptions Businesses may be located along the pipe ram alignment. The businesses near the shafts will be most affected by the work. Shafts should be located where they will not block driveways or sidewalks and prevent customers from reaching the adjacent businesses. The owner of the project should initiate a public outreach program to keep businesses and residents along the project alignment informed of the project's activities and schedule.

Visual Impacts Equipment including cranes will be staged around the portal/shaft locations and will be visible for the duration of construction.

Potential Settlement of Existing Facilities The use of pipe ramming typically results in minimal settlement or heave. The pipe ramming process does introduce energy into the ground, which can cause loose materials to consolidate or densify, resulting in ground settlement. Existing utilities and facilities can be affected by any settlement that occurs. Preconstruction and postconstruction surveys should be incorporated into the contract documents to protect from claims unrelated to construction.

Utilities Underground and Overhead The pipe ramming method is used to install casings, typically within the road or railroad's rights-of-way. Other utilities will be in the same rights-of-way. Portals/shafts installed to gain access to the pipe ram horizon may encounter these other utilities. Utilities in conflict with the pipe ram will need to be relocated. The pipe ramming method involves the construction of portals/shafts and the lowering of equipment to the pipe ram horizon. As a result, cranes are needed during the shaft construction and casing and carrier pipe installation. The location of overhead wires relative to a crane's boom and swing needs to comply with OSHA regulations.

Disposal of Spoils and Groundwater The pipe ramming process generates spoils during cleaning out of the installed casing. This material must be collected and disposed of. The water content of the material will dictate whether the material must be hauled off site in sealed trucks. Construction water, groundwater, and surface runoff may accumulate in the shafts and need to be pumped to the surface. It is sometimes possible to obtain a permit to discharge water collected from the site into the local storm or sanitary sewer system. The volume of material generated from the casing excavation is significantly less than that generated by open-cut construction. The volume of spoils is limited to that produced by the shaft and casing installation.

Bypass Pumping When replacing culverts or surface drains, bypass pumping may be required to collect and pump drainage water during pipe ramming. When consuming a culvert or pipe, the flow of water through an existing structure can occur when the hammer is not attached.

Public Events and Construction Moratoriums Public events and construction moratoriums can occur at any time of the year. Most are well planned and prepared in advance. Typically, a local government will maintain a list of upcoming events and construction moratoriums. If a public event or construction moratorium is planned for the period when the project is under way, the timing of the contract and the construction should be adjusted to minimize effects on the project.

3.3 PRELIMINARY GEOTECHNICAL CONSIDERATIONS

In the planning phase of the project, a preliminary geotechnical investigation should be performed to identify the general and any special subsurface conditions. The extent of the investigation required will vary, depending on known local conditions.

In areas of widespread uniform ground formations, it may be possible to use existing data to identify the ground types likely to be encountered. In areas of mixed, varying, or unknown existing conditions, a preliminary investigation should be undertaken to identify the general types of geotechnical conditions. A second function of the preliminary geotechnical investigation is to identify environmental concerns.

Initial site investigation work, also known as a *desktop study* (because much of the data can be gathered at the engineer's desk through phone calls and internet searches), is typically conducted during the planning stage of a project. It should include a site visit and a review of readily available sources of additional information concerning the site. The sources of additional information include geologic and topographic maps; aerial photographs; past geotechnical, environmental, or soil survey reports; construction case histories; professional papers about the area; historical documents concerning prior site usage; and discussions with local building officials and local contractors. In general, it is prudent to find out as much as possible about the site before planning and implementing further subsurface investigations.

A qualified engineer familiar with the alignment corridor should determine the extent of the preliminary geotechnical investigation. Preliminary geotechnical borings may be needed during the planning phase to understand alignment alternatives and fill information gaps. The geotechnical borings should be deep enough to bracket conditions that, if present, would affect construction. Recommendations for a complete geotechnical investigation, with borehole spacing and depths, are discussed in greater detail in Chapter 4.

3.4 PROJECT LAYOUT

3.4.1 Layout for Efficient Pipe Ram

Pipe ramming projects should be laid out with attention to the benefits and constraints unique to this construction process. The following recommendations address some of these concerns:

- Use standard casing ODs and wall thicknesses. Using nonstandard casing ODs may require custom-made adaptors for those specific casing diameters. Fabrication of new adaptors requires additional cost and time to mobilize to the construction site. Thinner wall thicknesses could jeopardize the success of the installation.
- Optimize crossing lengths consistent with casing length, casing size, pipe ramming equipment used, and geotechnical conditions.
- Evaluate casing diameter and length, different depths, carrier pipe sizes, and gradients. Increasing the casing diameter and length

increases the cost of material, the size of the support equipment needed, and the spoils and backfill quantities. The cost of constructing a deeper shaft adds cost to the project associated with increased excavation quantities, shaft support, and overall construction time.

Shaft locations are typically established as the end points of the crossing based on available workspace. The shaft locations can serve as maintenance access points/manhole locations, changes in direction, high and low points of a pressure line, and permanent structures located along the pipeline. Local operation and maintenance practices may require additional appurtenances. Local agency or owner constraints may need to be reevaluated based on the crossing length. If practical, the engineer may need to reduce the crossing length to accommodate maintenance or constraints.

Locations of shafts should be considered with respect to their effects on surface activities. Chapter 5 provides more discussion on shafts and shaft designs.

3.4.2 Alignment and Distance Considerations

Casing installations that use pipe ramming consist of straight drives by design. The intent is for the slope (grade) to remain constant for the length of the drive. During installation, if casing ends are not square and properly aligned or ground conditions change within the ram envelope, then deviations from the design line and grade can occur.

In general, the vertical alignment is established by the requirements of the project. Consider raising or lowering the casing profile to find uniform ground conditions and avoid unfavorable geology. When evaluating alignment alternatives, consider the effect on adjacent sensitive structures.

The length and diameter of a pipe ram installation should be evaluated relative to the risk of completing the drive. Risks to consider include exceeding the capacity of the casing or hammer, ramming the casing off the desired line and grade, and hitting buried objects.

3.4.3 Effects on Adjacent Structures

Adjacent structures should be identified, and the effect of pipeline installation on the structure should be determined. Adjacent structures include embankment slopes, road surfaces, rails, aboveground buildings, substructures, vaults, and utilities. The evaluation should include the horizontal and vertical position relative to the pipe ram. The zone of influence of both the casing installation and working shafts/portal faces should be identified. The age and condition of adjacent structures and utilities should be considered along with the materials of their construction, their construction history, their sensitivity to ground movement, and their foundation types.

With proper planning, pipe rammed casings typically have minimal or no impact on structures or utilities adjacent to the alignment. Monitoring of the casing advancement as compared to the bulked volume in the casing along with surface monitoring may be helpful in evaluating settlement and heave during installation. The bulked volume is the increased volume as a result of excavation of the in-place earth.

A survey of adjacent structures should be undertaken prior to construction. The survey should include photographs and videos of existing structures, including any visible cracks, settlement, or heave in existing ground surface. When very sensitive or high-risk structures are within a specific zone of influence, the installation and use of geotechnical instrumentation during construction may be required to provide early detection of ground movement so that corrective action can be implemented (Chapter 6).

With respect to existing utilities, their material construction and age should be considered. For example, older pipe may collapse or fail if the ground surrounding the casing moves. Older pipe with unrestrained joints might move, and its joints might open. Welded steel, fused plastic pipe, electrical distribution cables, and communication cables can usually tolerate some movement.

For planning purposes, the zone of influence is initially considered to be a 1:1 (horizontal:vertical) slope for most geological conditions, but the ratio might be higher or lower. For example, loose granular soil or very soft clays may have a zone of influence of 2:1. For very dense or stiff soils, the zone of influence may be reduced to 0.5:1. The engineer of the trenchless project should consult an experienced geotechnical engineer to determine the zone of influence based on the project's geology and anticipated ground behavior during the trenchless installation.

3.5 COST CONSIDERATIONS

Cost is an important consideration when comparing construction methods. Direct costs should not be the only consideration. Pipe ramming can be a more economical choice than traditional open-cut construction when all costs are included. The lower direct costs that increase the competitiveness of pipe ramming include less excavated volume, minimal surface disruption, less traffic control, and reduced project restoration.

3.5.1 Direct Cost of the Pipe Ramming

The direct cost of the pipe ramming method is influenced or controlled by several factors, which include the following:

- Geotechnical conditions and whether they are well documented for the bidders;

- Casing diameter and wall thickness;
- Buried objects along the alignment;
- Groundwater;
- Number of competing projects and number of qualified bidders;
- Labor conditions;
- Complexity of the project;
- Grade requirements;
- Location;
- Depth, size, and complexity of shafts;
- Total footage of installed casing;
- Mobilization, setup, and moving costs;
- Makeup time;
- Spoils removal and disposal;
- Effect of the pipe ram on existing utilities and structures;
- Special requirements and/or restrictions;
- Project owner's reputation; and
- Time of year/weather patterns.

Geotechnical Conditions Pipe ramming is only used in soils. Costs will potentially increase as the ground conditions becomes harder and denser because there is more resistance and larger hammers are needed to drive the casing into the ground. Squeezing ground increases the probability of becoming stuck in the ground. Gravels/cobbles/boulders increase the probability of stoppage, loss of grade, and damage to the leading edge/shoe or casing. Both of these conditions increase risk and cost to the project. If there is a lack of geotechnical information, bidders may include contingency costs in their proposals that will increase the contract price. The anticipated geotechnical conditions should be clearly stated in the bid documents to avoid these contingency costs.

Casing Diameter and Wall Thickness In general, large-diameter pipe rams cost more because they require larger and additional equipment, generate more spoils, require larger diameter casing with a thicker wall, and take longer to weld. As the wall thickness increases, the cost and weight of the casing increase. Because larger diameter casings excavate more ground per foot of the installation, such excavations require larger spoils removal equipment.

Buried Objects along the Alignment Hitting one or more buried objects can slow or stop the forward progress of the pipe ram and cause delays. Delays increase construction time, which in turn increases costs. When buried objects are encountered, additional time is needed for removal of the spoils, entry into the casing, and removal/breaking up of the buried object(s). Other options could include drilling the buried objects from the surface or the use of alternative tunneling methods from the opposite direction. Even buried objects that do not stop the rammed casing can deflect the installation away from its design alignment and grade, damage the casing, or increase the difficulty of spoil removal.

Groundwater Pipe ramming can be accomplished below the groundwater but this increases risk, and cost increases with depth. When groundwater is encountered the ground can flow into the face, carrying soil particles with it, and lead to possible voids in the ground above the casing. Should groundwater be present, this may require the use of special equipment, dewatering, more attention to shaft construction and watertightness, and sealing details. Costs of dewatering, pumping, and disposal of nuisance water increase with increasing shaft depth. Costs may also increase because of loss of production.

Number of Competing Projects and Number of Qualified Bidders The number of projects under construction, recently awarded, and in the bidding phase affect the backlog of the contractors. Contractors that have no backlog or a minimal backlog may set their bid prices lower just to have work for their employees. As contractors' backlogs increase, they can be more selective about which projects to bid and the prices they set for the work. The level of a project's risk, whether real or perceived, is critical to a contractor's interest in bidding. The number of qualified bidders also affects costs. Increased competition tends to reduce bid prices.

Labor Conditions The location of the project, the number of concurrent projects within a geographic area, and whether the area is subject to union rates or prevailing wage rates influence the number of workers and their costs.

Complexity of the Project As the complexity of the project increases, so do the project's risks and pricing. Increased complexity may reduce the number of bidders and drive costs up.

Grade Requirements When pilot tube guidance is used for grade control, it adds complexity and may add costs to production. If the contractor elects to use a smaller casing diameter and handle fewer excavated spoils during the construction process, the cost is less. If the contractor elects to increase the casing diameter, it allows greater flexibility in positioning the final carrier pipe, but it increases the cost of the casing installation.

Location If a project is located close to any established contractors for pipe ramming, the mobilization costs may be lower. If a pipe ramming contractor is on a local project, then mobilization costs are lower for other projects in the area if similar equipment can be used.

Depth, Size, and Complexity of Shafts The cost of a shaft incrementally increases with its depth and size, presence of groundwater, issues related to ground conditions, and complexity of site access. Shaft shape may also increase cost. Going from a circular to a rectangular shape has the additional cost of the internal support system (walers, struts, and bracing) that will be needed. In larger diameter shafts, the increased volume of excavated ground for a circular shaft may offset the cost of the internal support systems for a rectangular shaft.

Total Footage of Installed Casing As the total footage of a project increases, the unit cost per foot often decreases (economy of scale). The first crossing requires mobilization of equipment, construction of two shafts, and setup of all equipment. The second crossing on the same project will have a lower unit cost than the first crossing, because the costs of the mobilization are now spread across the two crossings. Once the equipment for a pipe ram is on site, and ground conditions remain relatively constant, the incremental unit cost for the pipe ram tends to decrease because the initial costs are spread over more footage, and the crew tends to become more productive through repetition of tasks.

Mobilization, Setup, and Moving Costs The major costs of a pipe ram are the equipment mobilization, initial setup, and total number of equipment moves. Mobilization costs include preparation of the equipment and transport of the equipment to the job site. Setting up all of the equipment and spoils removal system can take one day or more, depending on the complexity of the project. To move all the equipment to a new crossing location takes time. If two or more crossings are possible from the same launching shaft, preparing and turning around the equipment for the additional drives can be done in less time. It is most economical to design for crossings that are consistent with the equipment's capability, that minimize the number of required equipment setups, and that maximize the number of shafts from which multiple crossings can be executed.

Makeup Time One component of the advance rate is the makeup time, which is crew dependent. Makeup time is the time it takes to detach the hammer, install a new casing section, join the casing sections together, reattach the hammer, and restart the operations. It also includes the welding time. Time can be saved by using push-on type joints, but such joints can cost more to fabricate. When shorter pipe lengths are used, the number of casing sections increases. Longer casing sections require fewer joints but require construction of longer launching shafts. The cost of makeup time must be evaluated against the increased size of the launching shaft and the project's longer duration. Makeup time is independent of the advance rate.

Spoils Removal and Disposal Spoils from the installed casing require handling—including removal from the casing, removal from shaft, and disposal. Costs will increase with increased casing size and length. All spoils generated at the site require disposal. Increasing the distance to the final disposal site increases cost. Contaminated ground increases the cost of hauling and disposal.

Effect of the Pipe Ram on Existing Utilities and Structures The condition and location of the existing utilities or structures may affect cost and need to be assessed. Utilities may require relocation, replacement, or incorporation of protective measures. Additional cost may be needed to help locate utilities and structures and mitigate the effects from the pipe ram. Sensitive structures may require protective measures.

Special Requirements and/or Restrictions Special requirements imposed on a project increase the project's costs. Owners and engineers should work with third parties to minimize such special requirements. Typical special requirements or restrictions may include the following:

- Limits on work spaces;
- Limits on work hours;
- Clearance of road surfaces during peak commute hours;
- Limits on truck trips, casing and carrier pipe delivery, or spoils removal;
- Limits on noise or monitoring of noise; and
- Dust control and frequency of road cleaning.

Project Owner's Reputation An owner may have a "good" or "bad" reputation regarding contracting requirements and risk sharing. This can increase or decrease a contractor's contingencies on a project and affect a contractor's willingness to bid the work.

Time of Year/Weather Patterns Many pipe ramming projects involve culvert replacements that need stream controls or bypassing to construct the culvert. Work windows can be a function of the time of year or weather patterns that allow the construction. Delays can occur with extreme weather during construction or with permitting work windows. Any delays to a project will increase costs.

3.5.2 Indirect Social Effect Costs

The indirect social effect costs of a construction method should be factored into the construction method selection. Processes to incorporate these costs have been documented in Boyce and Bried (1994, 1998), Vickridge et al. (1992), and Sterling (1994). When alignments are through areas with heavy traffic, the indirect cost to the public may be the overriding consideration in choosing the pipe ramming method. The project affects an area from the shaft construction, through casing and carrier pipe installation, and site restoration. Informing the potentially affected parties about the duration and impacts before construction can reduce the costs of mitigating indirect social impacts during construction or prevent social impacts from becoming direct costs.

The cost of traffic disruptions is one indirect cost that is frequently considered. The engineer should use both the planned pipeline route and the proposed detour routes, if applicable, to determine the number of vehicles affected by construction.

Indirect or social effects include the following:

- Vehicle traffic and pedestrian disruption,
- Disruption to business activities,
- Loss of tax revenue,
- Disruption to residences,
- Carbon footprint,
- Noise,
- Dust,
- Odors,
- Restoration of pavements, and
- Public safety.

Vehicle Traffic and Pedestrian Disruption When large numbers of vehicles or pedestrians are slowed by pipeline construction in a roadway, the time lost by the individuals is considered a social effect. There is a social expense in terms of people's wasted time, which increases the anger of motorists and the potential for accidents. During a construction project, an owner or municipality is likely to receive and will need to deal with more complaints from the public. Pipe ramming minimizes the disruptions by installing the pipeline as a crossing under roadways.

Disruption to Business Activities Depending on the setup of the workspaces, access to businesses may be temporarily disrupted. Pipe ramming projects are typically positioned perpendicular to the roadway. Although parking spaces in front of stores may be needed during construction, access to the businesses and store fronts should remain open. Signage may be needed to let pedestrians and drivers know that businesses have remained open. Shafts should be positioned so as not to block the driveways of parking lots used or needed by the businesses.

Loss of Tax Revenue Loss of tax revenue, although considered by some to be an indirect social cost, actually is not, and therefore it should not be considered. Consumers buy products when and where they want or need. If access is restricted to a particular store, then the consumer will simply purchase the product at a different location. The amount of tax revenue generated and collected for a government entity is independent of where a product is purchased.

Disruption to Residences With any construction, there are disruptions to the public. Trenchless methods, including pipe ramming, are less disruptive to the surface than open-cut construction. In general, the effect related to social costs has to do with the loss of travel time for pedestrians and vehicles, as discussed previously. Pipe ramming generates noise, which must be addressed. Complaints concerning the amount of disruption and noise to a specific neighborhood typically are a function of the influence the residents have with the local government.

Carbon Footprint The amount of carbon dioxide generated by a project is now a consideration. As a result, programs exist to determine a project's carbon footprint. Pipe ramming uses fewer surface vehicles, creates less excavated material, requires less material handling, and historically has a lower volume of carbon emissions than open-cut construction.

Noise Noise will be generated by pipe ramming. The issue is how much noise is created above normal background noise. A constant noise is generated by the air compressor. The loudest noise generated by pipe ramming comes from the anvil striking the casing. The noise generated only occurs when the casing is being advanced in the ground. Periods of hammer striking noise are followed by longer periods of minimal noise when the new casing is being added to the installed casing. Local ordinances may affect whether work can be completed during off-hours. Noise reduction mitigation measures are a direct cost to the contractor.

Dust Dust is generated because of the removal of spoils and the movement of equipment. Dust is mitigated by the use of water trucks, covered dump trucks, and tracking mats. Fugitive dust can be cleaned with street sweepers. Dust reduction measures are a direct cost to the contractor.

Odors Odors can be generated from equipment (fumes). In general, these odors are limited. In some cases, excavations for shafts or the removal of spoils from the casing in organic or contaminated ground may generate odors that can be offensive to the public. In areas where submerged marine clays may be encountered, disturbance of the marine clay may release odors from decayed marine life. These odors may be difficult to mitigate during construction but can be avoided with proper planning of the alignment.

Restoration of Pavements Municipalities as well as regional and state agencies regularly renew pavements. These restoration projects usually consist of full-depth reconstruction, at a significant cost to these agencies. To maintain the structural integrity of the pavement as a whole, these agencies may require full sectional reconstruction of the pavement rather than simply patching the immediate area of disturbance. There is a direct cost to the contractor and, ultimately, to the project sponsor when full sectional reconstruction is required. There is also lost cost to the agency that expended monies to replace the pavement in the first place. Coordination, when possible, with the agency responsible for pavement maintenance should be conducted to adjust schedules and cost. In other words, project work should be done before any planned pavement replacement. In cases where older pavement is concerned, incisions caused by shaft excavations and by relocation of associated utilities may reduce the structural integrity and life expectancy of the pavement. This disruption may require expenditures of monies sooner than expected by the agency responsible for pavement maintenance.

Public Safety Surface support equipment for the construction will have a significant footprint at the launching shafts. The footprint may affect pedestrian use of sidewalks, reduce travel lane widths, restrict emergency vehicle access, or block access to hydrants or buildings in ways that affect public safety. Coordination with local emergency services should be conducted during the design and construction of the project. There is an associated cost for these additional services, which is typically borne by the local taxpayers but may be passed through to the contractor.

3.5.3 Environmental Costs

The effect of construction through contaminated ground and groundwater should be taken into account when potential contamination has been identified during the investigation process. A contingency plan that identifies methods for handling contamination may be required as part of the contract documents when construction will be through potentially contaminated areas. When possible, shafts should be located outside the contaminated area or constructed using methods designed to minimize the migration of contaminated ground and groundwater. Using pipe ramming below the level of the contaminated material can significantly reduce or eliminate the problems associated with installing a pipeline through these areas.

3.5.4 Risk Analysis

A risk register should be started for the project during the planning phase to help reduce overall project costs. The register is a living document and should be updated throughout the project's life. The engineer, with input from the owner, should identify and analyze the risks associated with viable construction methods. It may be helpful in some cases to conduct an alternative analysis of pipe ramming and other trenchless methods that accounts for both initial and long-term direct costs, indirect costs, and the impact of construction on third parties.

The process for developing a risk register includes the following activities (Ross 2012):

- Hold a brainstorming session with key members of the design team and the owner to identify project hazard and risk scenarios for each construction method proposed,
- Group the hazard and risk scenarios into related categories (e.g., shafts, trenchless methods, public outreach, permitting, property acquisition, safety, contractual issues),
- Identify the causes of each identified hazard or risk scenario,
- Determine the probability or likelihood that a specific hazard or risk scenario will occur,
- Determine the potential effect of the occurrence in terms of relative cost and/or scheduling delay to the project,
- Rate the hazard and risk scenarios on the basis of their likelihood and severity,
- Develop mitigation measures to reduce the probability and/or severity of hazard or risk scenarios,
- Rerate the hazard or risk scenarios with mitigation measures incorporated, and
- Update the risk register with new identified hazard or risk scenarios as the project advances in design and into construction.

As an example, the severity of the effect and the probability of a risk are both ranked from 1 to 5. Each risk ends up with a ranking between 1 and 25 (Table 3-1). Risks ranked in the red zone should be mitigated during design by reducing the severity or probability of the risk, thereby moving the risk into the yellow or green zone. Risks ranked in the yellow zone should be mitigated during design by reducing the severity or probability of the risk, thereby moving the risk into the green zone, to the extent possible.

The risk register becomes a project management tool for making sure that mitigation measures, once identified, have been incorporated into the design and construction.

Table 3-1. Risk Matrix with Severity of Impact and Probability of Occurrence.

Impact \ Probability	Rare (1)	Unlikely (2)	Possible (3)	Likely (4)	Most likely (5)
Very high (5)	5	10	15	20	25
High (4)	4	8	12	16	20
Moderate (3)	3	6	9	12	15
Low (2)	2	4	6	8	10
Very low (1)	1	2	3	4	5

Risks may exist for one or more drives on a trenchless project that could prevent them from being completed successfully. The effect of failing to complete a drive should be considered. A contingency plan should be prepared, identifying at least one alternative construction method.

Typical risks are evaluated based on cost, schedule, safety, and environment. The risk can be included in different categories, such as ground conditions, permitting, contracting, and shaft construction.

3.5.5 Contingency Costs

There are unknown or unforeseen conditions associated with any underground project. Contingency costs should be identified and included in any construction cost budgets prepared by the owner. Despite completion of a risk register, some risks cannot be eliminated or reduced by design or construction mitigations alone. To accommodate the costs associated with these risks, the owner should consider adding contingencies to the construction cost budget. These contingency costs can be held outside the construction contract or included as allowances within the bid schedule. The allowances are fixed dollar amounts, set by the owner, to cover additional items for the project that could not be quantified at the time of the bid. For example, allowances can be included for disposal of contaminated ground.

When owners show their willingness to identify and share project risks, the bidding contractors are more likely to include fewer contingencies in their bids.

REFERENCES

ASCE. 1998. *Pipeline route selection for rural and cross-country pipelines.* MOP 46. Reston, VA: ASCE.

ASCE. 2002. *Standard guidelines for the collection and depiction of existing subsurface utility data.* CI/ASCE 38-02. Reston, VA: ASCE.

ASCE. 2019. "United Nation's sustainable development goals." ASCE Policy Statement 517. Accessed April 3, 2020. https://www.asce.org/issues-and-advocacy/public-policy/policy-statement-517---united-nation-s-sustainable-development-goals/.

Boyce, G. M., and E. M. Bried. 1994. "Benefit-cost analysis of microtunneling in an urban area." In *Proc., North American No-Dig 1994, Dallas,Texas,* Paper A1. Arlington, VA: North American Society for Trenchless Technology (NASTT).

Boyce, G. M., and E. M. Bried. 1998. "Social cost accounting for trenchless projects." In *Proc., North American No-Dig 1998.* 3–12. Arlington, VA: North American Society for Trenchless Technology (NASTT).

Common Ground Alliance. 2019. *Best practices.* Alexandria, VA: Common Ground Alliance. http://commongroundalliance.com/best-practices-guide.

Ross, R. S. 2012. *Guide for conducting risk assessments.* Special Publication 800-30 Rev. 1. Gaithersburg, MD: National Institute of Standards and Technology (NIST).

Sterling, R. L. 1994. *Indirect costs of utility placement and repair beneath streets.* Rep. No. MN/RC-94/20. St. Paul, MN: Minnesota Department of Transportation.

Stuedlein, A. W., and T. Meskele. 2014. "Drivability of an instrumented 2440-mm diameter rammed pipe." In *Proc., North American No-Dig 2014, Orlando, Florida.* Arlington, VA: North American Society for Trenchless Technology (NASTT).

Vickridge, I., D. J. Ling, and G. F. Read. 1992. "Evaluating the social costs and setting the charges for road space occupation." In *Proc., North American No-Dig, Washington, DC,* Paper B2. Arlington, VA: North American Society for Trenchless Technology (NASTT).

CHAPTER 4
DETAILED SITE INVESTIGATIONS

4.1 GENERAL

This chapter addresses site investigations that are important for planning, detailed design, construction, and resiliency of pipe ramming projects. These site investigations may include the following:

- Geotechnical assessment,
- Utility surveys (both aerial and subsurface),
- Traffic flow and access for vehicles and pedestrians,
- Environmental conditions,
- Flood zones, and
- Seismic conditions.

This chapter also provides guidance for determining whether the pipe ramming method is appropriate based on identified or anticipated ground conditions (Section 4.2.4).

It is not the intent of this chapter to provide site investigation guidance for all conceivable pipe ramming projects. All projects are different and require site-specific investigations and input from experts within the appropriate fields.

4.2 GEOTECHNICAL ASSESSMENT

It is important that ground and groundwater conditions and their range of variability be reliably established as early as possible during planning to determine the feasibility of implementing pipe ramming. If pipe ramming is determined to be feasible, the information obtained will be used to design and construct the project. Problematic ground conditions can reduce the feasibility of, or eliminate, pipe ramming as an option for construction. Problematic ground conditions include ground types that are not compressible. Typical incompressible ground includes rock, cemented soils, and very stiff or very dense formations with firm behavior. If the anticipated largest particle size in the formation approximates the casing's smallest diameter, then the pipe ram would become in all practicality, full face rock and not feasible. Pipe ramming through rock—whether soft, fractured, weathered, or massive—is not feasible and must be avoided.

4.2.1 Site Investigation

A common and accepted practice is to conduct the site investigation in phases. Each subsequent phase of the study addresses information gaps identified in the prior phase of the investigation and provides information needed to plan the next phase of study.

In general, the first phase is a desktop study. Subsequent phases may consist of test s, supplemented with one or more test pits and/or observation wells with samples obtained for laboratory testing purposes. For a smaller project, this may be the primary design-level investigation.

On completion of this work, it is then possible to establish a working model for the site including ground properties, stratigraphy, and groundwater conditions. Use of observation wells will allow the collection of groundwater elevation information to assess the fluctuation of groundwater levels. Groundwater samples obtained from the wells can also be analyzed for water quality and corrosivity if the wells are installed in accordance with current environmental procedures.

Supplemental phases of investigation represent an opportunity to test the validity of the subsurface model. Depending on the outcome of this work and the importance of specific conclusions relative to subsurface conditions, it could be prudent to conduct additional phases of more detailed investigations to resolve remaining ground condition uncertainties. For additional discussion on conducting geotechnical investigations, refer to FHWA Report No. FHWA-IF-02-034 (Sabatini et al. 2002). For additional discussion of the phased approach for underground projects, refer to Parker (2004).

Desktop Study The desktop study typically consists of a review of existing information, which may include

- Geologic and soil survey maps;
- Local or regional groundwater (and surface water) conditions and historic highs and lows;

- Current and historical aerial photography;
- Historical maps and photographs;
- Existing or historical subsurface information, which may include boring logs of previous subsurface investigations;
- Construction records from previous nearby projects or projects in similar subsurface conditions; and
- Potential for contamination, including review of state and private website tracking tools and record searches as well as aerial photography, mapping, and photography.

Depending on the scope of the project, the results of the desktop study may be published as a preliminary geotechnical report or memorandum or included in a preliminary design report. For some projects, the results of the desktop study may be used to plan the subsequent subsurface investigations and may be included with the primary geotechnical reporting.

Subsurface Investigation Methods Initial site investigation work, which is typically conducted during the planning or preliminary design stage of a project, should include a site visit and a review of material gathered during the desktop study.

The most common form of subsurface investigation is the sampled test boring. The method of drilling and sampling for the test boring may vary based on the location, anticipated ground conditions, and site access. The method involves advancing a small-diameter vertical borehole and collecting samples of the ground encountered. Important first decisions are to determine what method of test boring, how many test borings to drill, and to what depth. Many guidelines include suggestions on the number of test borings per foot of pipeline. Test borings should be located at launching and receiving shaft locations unless suitable historical data exist and should extend deeper than the expected pipe ram depth plus the depth for the shaft foundation and embedment. Depending on the consistency of subsurface conditions, test borings at the shaft locations may be sufficient for shorter, less than 200 ft (60 m), pipe ramming installations. Additional borings are appropriate for pipe ramming drives longer than 200 ft (60 m).

The final determination of the number and depth of borings should be based on the sensitivity and complexity of the project, the anticipated risks, site access, and the nature and complexity of the ground conditions. The scope and results of the subsurface investigation program should be reviewed by geotechnical personnel experienced with the local geologic conditions and familiar with pipe ramming. The magnitude of the subsurface investigation program should reflect the risk associated with the project. The more difficult and costly the proposed installation, the more potential effects to third parties; the more complex the ground conditions, the more extensive the subsurface investigation program should be.

At some sites, it may be prudent to supplement the test boring program with other forms of subsurface investigation such as test pits, cone penetration tests (CPTs), rotosonic borings, large-diameter drill holes, and geophysical methods. Test pits, rotosonic borings, and large-diameter auger holes are useful methods of subsurface investigation for identifying the presence and magnitude of gravels, cobbles, and boulders; where subsurface conditions are extremely variable; or in urban settings in which areas of undocumented fill may exist. These methods may also provide direct evidence of the type and distribution of buried objects that exist in the ground and reveal important facts about possible ground behavior characteristics such as sloughing, squeezing, and groundwater inflows. CPTs involve pushing a piezocone probe into the ground and can be a cost-effective method of obtaining a continuous record of subsurface conditions. CPTs should always be "calibrated" at the site with one or more coupled soil borings, and their use is contingent on soil density/hardness and CPT rig capacity.

Test borings at shaft locations should extend sufficiently below the shaft bottom to identify changes in the strata below both the shaft and pipe ram alignment that could affect the construction of the shaft and to provide information for design of the ground support system. The additional boring depth provides geotechnical information for design of the shaft foundation and invert slab, if applicable. Explorations should extend at least 10 ft (3 m) below the tip elevation of the proposed shaft ground support systems. Test borings along the pipe ram alignment should extend approximately 10 ft (3 m) or two casing diameters below proposed casing invert, whichever is deeper. If soft, compressible soils are present at or near the invert of the proposed pipe ram, some of the borings should be extended into firm ground, if possible.

For most geotechnical test borings, soil samples are obtained using the standard penetration test (SPT), in accordance with ASTM D1586 (ASTM 2011). The SPT involves driving a hollow 2 in. (50 mm) outside diameter steel split spoon sampler 18 to 30 in. (450 to 760 mm) long and with a 1-3/8 in. (35 mm) inside diameter, using a 140 pound hammer, falling freely over a vertical distance of 30 in. (760 mm). The procedure involves driving the sampler either 18 or 24 in. (450 or 600 mm) and recording hammer blows in 6 in. (150 mm) increments. The sum of the blows recorded over the last driven 12 in. (300 mm) is defined as the N-value and provides a measure of the relative density or consistency of the soil. ASTM D1586 provides further information. For samplers and hammers that deviate from ASTM D1586 (e.g., ASTM D3550), there are correlations; however, the reader is cautioned that these field blow counts do not represent N-values. For additional information consult a geotechnical engineer.

Typically, split spoon samples should be taken at intervals no greater than 5 ft (1.5 m) and at changes in geologic strata. Continuous sampling within the anticipated pipe ram envelope is prudent to identify thin strata that could have significant effects on the pipe ram advancing. If classifications of soil and rock are the only tests required, then disturbed sampling procedures are acceptable (i.e., split spoon). If, however, laboratory strength and compressibility tests are needed, undisturbed samples may be required in lieu of split spoons.

If rock fragments are encountered within the sampler during the geotechnical investigations, steps should be taken to determine if the rock pieces are naturally occurring gravels or if the sampler broke up larger pieces of rock into fragments. Where there is a potential for gravels, cobbles, and boulders, large-diameter auger holes, rotosonic borings, test pits, HDD probes, or trenches can be useful in obtaining representative samples for grain-size analysis and determining the size and distribution of the gravels, cobbles, and boulders.

If bedrock is encountered, consider continuing with coring using triple-tube (e.g., HQ, NQ) core barrels to determine if the rock is a large boulder or bedrock. Coring is also a viable sampling method if oversized materials (e.g., cobbles, boulders) are encountered. Pipe ramming should be avoided if rock conditions are encountered.

Field tests that might be used include in situ vane testing, hand penetrometer testing, point load testing on gravel and rock pieces of cobbles and boulders, pressure meter testing, and in situ downhole permeability or pumping tests.

For project sites where obtaining vertical borings along the alignment is not possible, the proposed alignment can be explored using small-diameter HDD rigs. The progress and drilling efficiency of the HDD probes can be tracked. Areas of gravels, cobbles, boulders, or buried objects can be identified by the HDD probe. The HDD probe can be advanced multiple times within the proposed pipe ram envelope to evaluate the presence of challenging ground.

There are also many innovative and cost-effective methods of investigating the ground by geophysical techniques and other emerging methods. These investigations can be effective in helping to interpolate subsurface conditions between the test borings and locating the boundary between soil and rock. Although a complete listing of all possible geophysical exploration methods is beyond the scope of this section, some examples are

- Seismic refraction and reflection surveys,
- Ground penetrating radar, and
- Electrical resistivity studies.

These methods and downhole investigations have been shown to be effective in some types of subsurface profiles. In general, such applications should be used if they provide useful information at a nominal cost relative to the proposed design and construction requirements. The effectiveness of these methods may depend on the subsurface conditions, nearby sources of interference, and the expertise of the operator. An experienced geophysical professional should be consulted during planning of the geophysical program to ensure appropriate methods are used.

Other success factors for a subsurface investigation program include good quality exploration and sampling services and continuous on-site monitoring of the work by a trained professional representing the engineer. Thorough observation and documentation of field activities must be provided so the results of the field work can be accurately and correctly interpreted.

4.2.2 Characterization of Subsurface Materials

Certain characteristics are important to identify and quantify during the site investigation because they may change the effectiveness of the pipe ramming installation. Site characteristics that should be identified for pipe ram projects are

- Ground conditions (very soft to hard/very loose to very dense);
- Rock conditions;
- Gravels, cobbles, and boulders;
- Groundwater conditions;
- Potential buried objects;
- Contaminated or hazardous ground or groundwater; and
- Other considerations such as environmentally sensitive surface features, biological entities, and the potential for cemented soils/caliche.

Ground Conditions An evaluation of whether the soil will be readily penetrated is a critical component in assessment of the feasibility of the pipe ramming installation. Typically, the density or consistency of the soil can determine whether the ground can be penetrated by the leading edge of the casing. A detailed summary of pipe ramming applicability in various ground conditions is provided in Section 4.2.4.

For cohesionless soils (predominantly silts, sands, and gravels), important properties that should be identified include

- Grain-size distribution,
- Unit weight,
- Hydraulic conductivity,
- Density (typically in terms of SPT N-values), and
- Maximum clast size (i.e., gravels, cobbles, and boulders if present).

For cohesive soils (clays and silts), important properties that should be identified include:

- Moisture content,
- Atterberg limits (plasticity index and liquid limit),

- Unit weight,
- Shear strength,
- Compressibility,
- Consistency [typically in terms of unconfined compressive strength (UCS) or SPT N-values],
- Grain-size distribution sieve and hydrometer analysis, and
- Maximum clast size (i.e., gravels, cobbles, and boulders, if present).

Laboratory testing should be done in accordance with the appropriate ASTM standards.

The Tunnelman's ground classification system, first described by Terzaghi (1950) and refined by Heuer (1974), is designed to describe ground behaviors and their effect on larger, conventionally constructed soft ground tunnels. The system is also a useful tool for evaluation of soft ground conditions as they relate to pipe ramming. Descriptions of various ground behaviors, in terms of the Tunnelman's ground classification system and its effects on pipe ramming, are presented in Table 4-1. General ground behaviors include raveling, running, flowing, squeezing, and firm ground. See Section 4.2.4 on the applicability for pipe ramming and further discussion of the importance of the ground classification.

Rock Conditions The use of pipe ramming is not feasible in rock. Consider other trenchless methods that can excavate through rock conditions if anticipated. The elevation and variability of the rock strata relative to the pipe ram envelope should be determined. The boundary between soil and bedrock UCS is commonly described as 140 psi (20,000 psf) (Kulhawy et al. 1991), and installations close to or along soil/rock interfaces should be avoided.

Gravels, Cobbles, and Boulders The success of a pipe ram may depend on the diameter of the pipe ram as compared to the size, distribution, and strength of the gravels, cobbles, and boulders. The extent, frequency, size distribution, maximum size, and physical properties of gravels, cobbles, and boulders should be determined. Pipe ram drives should be designed to minimize the effect of gravels, cobbles, and boulders (i.e., oversized casing) or located to avoid gravels, cobbles, and boulders if possible. If these conditions cannot be avoided, pipe ramming typically becomes the trenchless method of choice, because the spoils in the casing can be removed and inspection of the cobbles and boulders at the leading edge is possible.

One way to document the frequency of the cobbles or boulders is to determine the cobble volume ratio (CVR) and the boulder volume ratio (BVR). The CVR and BVR are the ratio of the volume of cobbles or boulders to the excavation volume reported in terms of percentage (Hunt et al. 2013).

Table 4-1. Tunnelman's Ground Classification for Soils.

Classification		Behavior	Typical soil types
Firm		Heading can be advanced without initial support and final lining can be constructed before ground starts to move.	Loess above water table; hard clay, marl, cemented sand, and gravels when not highly overstressed.
Raveling	Slow raveling Fast raveling	Chunks or flakes of material begin to drop out of the arch or walls sometime after the ground has been exposed. This is due to loosening or to overstress and *brittle* fracture (ground separates or breaks along distinct surfaces, as opposed to squeezing ground). In fast raveling ground, the process starts within a few minutes, otherwise the ground is slow raveling.	Residual soils or sand with small amounts of binder may be fast raveling below the water table; slow raveling above. Stiff fissured clays may be slow or fast raveling, depending on degree of overstress.
Squeezing		Ground squeezes or extrudes plastically into tunnel, without visible fracturing or loss of continuity, and without perceptible increase in water content. Ductile, plastic yield and flow due to overstress.	Ground with low frictional strength. Rate of squeeze depends on degree of overstress. Occurs at shallow to medium depth in clay of very soft to medium consistency. Stiff to hard clay under high cover may move in combination with raveling at excavation surface and squeezing at depth behind surface.

Classification		Behavior	Typical soil types
Running	Cohesive running Running	Granular materials without cohesion are unstable at a slope greater than their angle of repose (±30° to 35°). When exposed at steeper slopes, they run like granulated sugar or dune sand until the slope flattens to the angle of repose.	Clean, dry granular materials. Apparent cohesion in moist sand or weak cementation in any granular soil may allow the material to stand for a brief period of raveling before it breaks down and runs. Such behavior is cohesive running.
Flowing		A mixture of soil and water flows into the tunnel like a viscous fluid. The material can enter the tunnel from the invert as well as from the face, crown, and walls, and can flow for great distances, completely filling the tunnel in some cases.	Below the water table in silt, sand, or gravels without enough clay content to give significant cohesion and plasticity. May also occur in highly sensitive clay when such material is disturbed.
Swelling		Ground absorbs water, increases in volume, and expands slowly into the tunnel.	Highly preconsolidated clay with a plasticity index in excess of about 30, usually containing significant percentages of montmorillonite clay.

Source: Adapted from Heuer (1974).

Groundwater Conditions The groundwater level in unconfined aquifers and piezometric levels in confined aquifers should be determined by installing observation wells and/or piezometers adjacent to shaft locations. Consider adding piezometers at intermediate points for longer drives. The hydraulic conductivity of water-bearing strata should also be determined.

Groundwater conditions will have a significant influence on ground behavior (Table 4-1), the viability of pipe ramming, launching and receiving shaft design, and construction. Pipe ramming in certain ground conditions can cause localized liquefaction and consolidation. Hydraulic conductivity of water-bearing strata can be estimated using grain-size correlations or measured by borehole permeability tests. Pump tests may be warranted for larger projects if highly permeable soils with significant groundwater control issues are anticipated during construction. Hydrologists that specialize in groundwater characterization for trenchless construction should be consulted if groundwater is considered a significant project risk.

Dewatering in the immediate vicinity of active pipe ramming operations can be an effective way to improve the ground and facilitate construction. However, dewatering may have unintended consequences such as increased friction on the casing, resulting in higher ramming forces because of increases in effective stress and loss of lubrication.

Potential Buried Objects The likelihood of buried objects, their nature, and relative sizes should be evaluated by the desktop study and site investigation. Pipe ramming can be stopped by buried objects in the path of the ram depending on the composition, orientation, and size of the buried object. Known objects should be avoided if possible. Examples of buried objects from past construction include foundations, old pile-supported structures, cisterns, slurry walls, drain fields, bulkheads, and wells. The potential for buried objects and an understanding of their nature should be determined early in the planning phase to minimize the risk of encountering them. To increase the likelihood of engulfing or breaking up the buried object, consider using a cutting shoe and enlarging the diameter of the casing.

Contaminated or Hazardous Ground or Groundwater Encountering soil and/or groundwater contamination has health and safety, cost, and schedule impacts on projects. Hazardous conditions can include naturally occurring hydrocarbons and asbestos. Other hazards can include naturally occurring gases such as methane and hydrogen sulfide. Careful planning, research, and execution may minimize these impacts during construction. Determination of the potential for encountering contaminants and hazardous substances should be completed during the planning phase. If contaminants are found, then determination of their nature and extent must be undertaken during the site investigation. Even if contaminants are not identified during the planning phase, the site investigations should screen for contaminants. One inexpensive approach is to sample the headspace above the samples with a photoionization detector (PID) and to record the readings. Consult an environmental professional for additional information regarding detecting the potential presence of contaminants in the ground and possible effects on the project.

Other Considerations Other important information to characterize includes environmentally sensitive surface features, biological entities, and the potential for cemented soils/caliche. Environmentally sensitive areas can include wetlands, riparian corridors, or environmentally protected areas such as a nesting habitat. In some areas of the country, environmentally sensitive areas are seasonal. It is important to understand noise, vibration, and construction windows for the project location.

Biological entities can be sensitive to pipe ramming. Examples include sensitive plants, roots, birds, small mammals, and marine animals.

Additional information about environmentally sensitive areas and contaminated soils is provided in Section 4.5.

Cemented soils and caliche are regionally specific soil concerns in arid environments. It is important to consult geologists or geotechnical engineers familiar with local geology to assess the potential for this risk.

4.2.3 Geotechnical Reports

All relevant subsurface data collected during the geotechnical desktop study and site investigations, professional interpretations thereof, and design and construction considerations should be summarized in project reports. The geotechnical data report (GDR) contains all the factual geotechnical information for the project, including explorations, laboratory and field testing results, and geophysical and historical geotechnical data. The GDR does not include interpretations or recommendations and should be included in the contract documents.

Geotechnical design memoranda and/or a geotechnical interpretive report (GIR) should be prepared to present summaries of the geotechnical data, interpretation of the data, earth pressures to be used for design, discussions of the expected behavior of the ground, and other geotechnical design recommendations such as dewatering, pipe ramming, shaft types, and support systems. Because the GIR and design memoranda are typically prepared prior to design, the information presented in these memoranda and reports with respect to the project may not be applicable to the contractor at the time of bid. These design documents should be disclosed to the bidders but are typically not provided to the bidders, nor should they be included with the contract documents.

Increasingly, another stand-alone report is also prepared, known as a geotechnical baseline report (GBR). The GBR is typically prepared with the owner's input as the design is being completed and serves as the definitive geotechnical baseline for use in the resolution of disagreements, disputes, or claims related to differing subsurface conditions. The GBR presents contractual interpretations of the data to be used for bidding and construction, as well as baseline expected behavior of the ground and other geotechnical construction considerations, such as appropriate shaft types and systems.

See *Geotechnical Baseline Reports for Construction: Suggested Guidelines* (ASCE Technical Committee for the Underground Technology Research Council 2007). The GBR should be included in the contract documents and prepared by a qualified geotechnical engineer or engineering geologist experienced in underground construction.

4.2.4 Applicability of Pipe Ramming Based on Subsurface Conditions and Classification

The information in Table 4-2 can be used to determine whether pipe ramming is an appropriate method for the identified or anticipated geotechnical conditions, or whether pipe ramming in general is a feasible method for the project.

Because pipe ramming is typically an open-faced trenchless method, it is not considered applicable for use in ground conditions that classify as flowing (according to terminology used in Table 4-1). Ground improvement (e.g., dewatering) can be implemented to change the ground classification, or modifications to the construction approach (e.g., constructed soil/sandbag plugs) can be used to reduce the risks associated with flowing ground. Similarly, squeezing ground can result in high ramming forces, which may exceed the capacity of the chosen equipment. Identification and mitigation of the risks associated with these ground types can be key to successful completion of a drive.

4.3 UTILITY SURVEYS

It is important that the locations of existing and abandoned utilities be reliably established as early as possible during planning to determine the feasibility of implementing pipe ramming. Existing utilities should be identified in an iterative manner whereby each level of survey builds on the prior study and identifies data gaps to be filled in a subsequent study. Existing utilities should be avoided, or the utility should be relocated before construction. Abandoned utilities should be identified and avoided or removed before construction.

All utility information (aerial and subsurface) should be collected in connection with the project alignment(s) and presented in accordance with ASCE 38 (2002). It is recommended that QLB and QLA utility data be obtained, especially if the utility expectancy is complex and/or risky. Note that using a *design ticket* from a state's on-call service can result in only QLC or QLD information.

During the planning and design phase, it may be possible to make changes to the pipe ram alignment and shaft locations to avoid conflict and the necessity for potentially costly and time-consuming utility relocations.

Table 4-2. Applicability of Pipe Ramming for Identified Geotechnical Conditions.

SPT N-value	Geotechnical condition	Applicability
0–1[a]	Very soft clays, silt, and organic deposits	Depends[b]
2–4	Soft clays, silt, and organic deposits or very loose sands	Yes
5–10	Medium stiff clays and silts or loose sands	Yes
11–30	Stiff to very stiff clays and silts or medium dense sands	Yes
11–30	Soil with gravels [1 to 3 in. (25 to 75 mm)]	Yes
11–30	Soils with occasional cobbles [3 to 12 in. (75 to 305 mm)], boulders (>12 in.)	Depends[c]
11–30	Soils with significant cobbles, boulders, and obstructions larger than 4 in. (100 mm)	Depends[c,d,e]
31–50	Hard clays or dense sands	Depends[e]
>50	Very hard clays, very dense sands, or rock	No

Note: In general, larger diameters have a better success rate in more complicated geotechnical conditions.

[a]Also referred to as *weight of hammer material.*

[b]The ground conditions are likely too weak to maintain line and grade.

[c]Likelihood of completion depends on diameter of the pipe as compared to the size of the buried object hit, whether the buried object can be engulfed or if the pipe hits the buried object.

[d]Pipe ram will require heavier walled casing, a robust cutting shoe, and may not achieve line and grade.

[e]Pipe ram may be able to penetrate a portion of the ground with this N-value but not for an extended distance.

If conflicts cannot be avoided, relocation plans for the utilities should be prepared. Relocation of the utilities can be done either in advance of the project or as part of the project and can be of either a temporary or permanent nature. The utility owner must be engaged as early as possible for input and review. It is beneficial to contact the local community for input and information. All known existing and abandoned utilities should be shown in the contract drawings. Information such as utility quality level, diameter, depth, pipe material type, owner, trenched cross section, and backfill material used (according to ASCE 38) should be identified, if available.

Utilities that will remain in place and are in proximity to the project may require protection and monitoring during construction. Evaluation of settlement and heave risks may be appropriate depending on type and age of the utility, clearances, and ground conditions. Real-time settlement monitoring and vibration monitoring of critical utilities may be required by the utility owner if the utility cannot be relocated.

4.4 TRAFFIC FLOW AND ACCESS FOR VEHICLES AND PEDESTRIANS

Traffic control measures taken or traffic control or calming devices installed to alert drivers or modify the overall flow of traffic should be planned and approved in advance with the agency responsible for maintaining unimpeded traffic flow and according to the Federal Highway Administration's *Manual on Uniform Traffic Control Devices* (MUTCD) (FHWA 2012).

If the proposed pipe ram alignment or shaft location affects traffic, it may be important to collect and evaluate traffic information. The traffic information can be used to determine the feasibility of implementing pipe ramming; to develop alignment alternatives; to plan shaft locations, staging areas, and drive lengths; and to develop means to minimize disruption to vehicles and pedestrians. As a general statement, pipe ramming is used as a crossing method, whereby the casing is installed perpendicular to the roadway as opposed to along the roadway. The extent of the traffic data requirements is dependent on the nature, complexity, and setting of the proposed construction. The presence of individual facilities with unique access demands (such as schools, hospitals, fire stations, post offices, distribution centers, and bus depots) must also be considered.

During planning and route selection the alternatives analysis should include the real costs and the social costs (see Chapter 3 for social cost discussion). Real cost impacts include design and construction costs. Design costs include preparation of maintenance of traffic plans. Construction costs can include traffic control and enforcement. Alternative traffic management strategies through work zones may have substantial project cost consequences. For example, if pipe ramming requires time restrictions, such as construction during off-peak hours only, the effective workday may be reduced significantly. Thus, the overall duration of project construction increases, substantially affecting project costs. Similarly, other requirements may influence the available work area during certain phases of construction, such as workspace limitations that may be caused by parallel or adjacent construction activities with potential increased project cost consequences.

For most projects, traffic impacts cannot be completely eliminated. Even when the route is fixed, the selection of shaft locations and workspaces can have a significant impact on traffic.

4.5 ENVIRONMENTAL CONDITIONS

Environmental conditions can be naturally occurring or the result of human activity. Some naturally occurring conditions are naturally occurring hydrocarbons and asbestos as well as active or dormant landslides, active erosional features, and corrosive soils. Some human-created conditions are places of historical and cultural significance as well as landfills and contaminated ground and groundwater. Environmentally sensitive areas can include

- Sources of water supply;
- Islands and island corridors;
- Beaches;
- Dunes;
- Wetlands and wetland transition areas;
- Breeding, nesting, and spawning areas;
- Migratory stopover areas;
- Wintering areas;
- Prime fishing areas;
- Migratory pathways;
- Water areas supporting submerged vegetation;
- Shellfish harvesting waters;
- Forest areas;
- Habitat for federal and state endangered or threatened plant and animal species;
- Federal and state wilderness areas; and
- Areas designated as wild, scenic, recreational, or developed recreational rivers.

Properly planned pipe ramming can reduce or eliminate the effect of construction on these environmental conditions by undercrossing the features or by scheduling construction during nonpeak hours or inside specific construction-time windows.

Existing site features that could be affected by shaft construction and pipe ramming operations should be identified during the site investigation. Historic buildings and environmentally sensitive areas usually require evaluation on a case-by-case basis to ensure protection.

Effects to the surface or the near surface environment can be reduced by locating launching and receiving shafts outside the sensitive area and pipe ramming under it. In this case, the effects are avoided by simply passing under the sensitive area. If the distance is too great for a single drive, then consider either (1) using a pilot tube for guidance, pipe ramming from two different directions, and meeting in the middle, or (2) using a different trenchless method.

For contaminated ground, if the affected area cannot be avoided during planning-level route selection, then the volume of contaminated spoils generated can be reduced by locating the shafts outside the contaminated area and then advancing the casing though the affected area. In this case, the volume of contaminated ground is reduced to the excavated casing volume. When excavating contaminated ground, always consider the total excavated volume (shafts and pipe rammed casing) in arriving at a final design.

Consult an environmental professional for assistance in development of contract requirements related to such issues as handling and disposing of the spoils, decontamination of equipment before and after construction, and personnel health and safety. The availability of contractors qualified to perform work at contaminated sites may be reduced.

4.6 FLOOD ZONES

The potential for flood hazards should be considered during design when identifying and selecting shaft locations and pipe ram alignments in low lying areas or areas prone to sudden floods, storm surges, and flash flood conditions. Designing for potential high water conditions can minimize flooding of shafts, reduce damage to pipe ramming equipment, prevent pollution from petroleum products, mitigate cost and schedule impacts, and protect personnel. In addition, locating work areas and shafts within designated flood zones may require additional permitting effort, which can delay design activities.

Assessment of flooding risk during planning should include review of information from FEMA, the National Oceanic and Atmospheric Administration (NOAA), and USACE. Such assessment should also include local knowledge to determine whether storm impacts are probable, along with expected seasonality and frequency of storm occurrence. In some locations, flooding may last for weeks and affect the project duration. The ground conditions may become saturated and can affect the shafts and ramming operation.

If flooding conditions are a significant risk, the design may need to consider the following:

- Design shafts to address the full hydrostatic pressure condition (i.e., increase pressure to the top of the shaft or plan to flood the shaft to eliminate unbalanced hydrostatic pressures).

- Design for fully saturated ground conditions.
- Temporarily raise the site grade, if permittable, to provide better access.
- Design for increased uplift that may occur during flooding or determine maximum groundwater level and method of monitoring that is allowed before flooding occurs.
- Increase pipe joint pressure ratings to accommodate installation and operational high water pressures due to flooding. Design may need to consider 10-, 50-, 100-, or even 500-year flood frequency events.
- Develop plans for evacuation and note that access roads may flood before the project site floods.
- Protect or remove sensitive equipment, oils, and fuels.
- Develop plans to protect the project site from flooding when not occupied, including weekends and holidays.
- Develop plans to monitor water levels in advance and during large rainstorm events.
- Design shafts, casings, and operations that prevent destabilization of flood control levees and embankments. Recognize that localized levee slope failures can occur during floods.
- Plan construction operations to avoid periods with a high potential for flooding (i.e., hurricane season, rainy season).
- Develop start-up plans after a flood event. Plans should include mobilization of pumps; disposal of water from the shaft; and cleanup of the surface, shaft walls, and equipment within the shaft.

In areas prone to flooding, the owner should consider including allowances or contingencies in the contract to compensate the contractor for this event, associated cleanup, and loss of time to the construction schedule. A major unexpected flood event may be considered a differing site condition. It is recommended that flood events be defined in reference to specific water surface elevation and recurrence intervals (e.g., 1 in 100 years) and the anticipated baseline conditions identified in the contract documents.

4.7 SEISMIC CONSIDERATIONS

In projects for which the rammed casing is the final carrier pipe, seismic factors (including liquefaction, embankment instability, ground shaking, fault offset, and related ground responses) should be considered. For temporary contractor designed structures such as launching and receiving shafts and for sacrificial pipe rammed casings, seismic considerations are less critical.

For permanent structures, it is also important to consider the anticipated postseismic event level of service and the consequences of failure for the pipeline or casing (i.e., flooding or gas release and fire). The seismic considerations include protection of personnel and minimization of damage from fault offset, ground shaking, landslide, liquefaction, lateral spreading, and tsunamis. Seismic experts should be consulted for design recommendations for permanent structures in seismic hazard areas.

In areas prone to seismic events, the engineer should consider the need for mitigation measures such as ground improvement, oversized casings around product pipes to allow fault offset/creep deformation to be accommodated by the casing, and placement of annular space (air or crushable material) between casing and product utility. The project could also be affected and delayed if an earthquake occurs during construction. For additional information on seismic design of pipelines, consult O'Rourke (2002) and the Technical Council on Lifeline Earthquake Engineering (TCLEE) monographs (https://ascelibrary.org/page/books/s-tclee).

REFERENCES

ASCE. 2002. *Standard guidelines for the collection and depiction of existing subsurface utility data.* CI/ASCE 38-02. Reston, VA: ASCE.

ASCE. 2007. Technical Committee for the Underground Technology Research Council. *Geotechnical baseline reports for construction: Suggested guidelines.* Reston, VA: ASCE.

ASTM International. 2011. *Standard test method for standard penetration test (SPT) and split-barrel sampling of soils.* ASTM D1586. West Conshohocken, PA: ASTM.

ASTM. 2017. *Standard practice for thick wall, ring-lined, split barrel, drive sampling of soils*, ASTM D3550. West Conshohocken, PA: ASTM.

FHWA (Federal Highway Administration). 2012. *Manual on uniform traffic control devices.* Adopted under 23 Code of Federal Regulations (CFR), Part 655, Subpart F. Accessed April 3, 2020. https://mutcd.fhwa.dot.gov/.

Heuer, R. 1974. "Important ground parameters in soft ground tunneling." In *Proc., Specialty Conference on Subsurface Exploration for Underground Excavation and Heavy Construction*, 41–55. Reston, VA: ASCE.

Hunt, S. W., D. E. DeNero, and A. J. Finney. 2013. "Microtunneling in gravel, cobbles and boulders." In *Proc., Rapid Excavation and Tunneling Conference (RETC)*, Washington, DC: SME.

Kulhawy, F. H., C. H. Trautmann, and T. D. O'Rourke. 1991. "The soil-rock boundary: What is it and where is it?" In *Detection of and Construction at the Soil/Rock Interface*, ASCE Geotechnical Special Publication 28, 1–15. Reston, VA: ASCE.

O'Rourke, M. J. 2002. "Buried pipelines," Chap. 23 in *Earthquake engineering handbook*, edited by Charles Scawthorn and Wai-Fah Chen. Boca Raton, FL: CRC Press.

Parker, H. W. 2004. "Planning and site investigation." In *Tunnelling: 1° Congresso Brasileiro de Túneis e Estruturas Subterrâneas Seminário Internacional,* South American Tunnelling. London: ISSMFE.

Sabatini, P. J., R. C. Bachus, P. W. Mayne, J. A. Schneider, and T. E. Zettler. 2002. *Evaluation of Soil and Rock Properties.* Geotechnical Engineering Circular No. 5, FHWA-IF-02-034. Washington, DC: FHWA.

Terzaghi, K. 1950. "Geologic aspects of soft ground tunneling," Chap. 11 in *Applied sedimentation*, edited by P. D. Trask. New York: Wiley.

CHAPTER 5
DESIGN

5.1 GENERAL

This chapter discusses design specific to construction with pipe ramming. In most pipeline construction projects, design and project conditions are summarized in contract documents that are prepared by the owner's engineer. These documents are the means of communicating project requirements to a contractor for pricing and must represent a constructible project, using industry standard practices associated with pipe ramming.

Design is a stepped process of planning, verification testing with site-specific data, mitigation, and documentation. As with any construction interacting with subsurface conditions, design success requires owner education by the engineer regarding site hazards, which is supported by limited subsurface data. The following is a brief summary of the design tasks:

- *Planning and permitting*: Define owner success factors and tolerances; identify possible impediments to the ramming process, access, ability of meeting owner requirements within acceptable tolerances (based on readily available data, local experience, and experience with pipe ramming construction); quantify and qualify planning cost and schedule; and develop methods for subsurface verification or characterization study.
- *Exploration*: Determine the ground truth of hazards and characterization of subsurface conditions.
- *Design*: Assess constructability, address permit conditions, and perform quantification for bidding.

- *Documentation*: Reports, bid documents, engineer's cost estimate, and schedule.
- *Design support during construction*: Respond to requests for information (RFIs), review and accept submittals, verify design assumptions, attend construction progress meetings, provide inspections or make site visits, recognize and respond to differing site conditions, and if necessary modify the design and identify acceptable resolutions to encountered conditions.
- *Constructability reviews.*

For trenchless projects, hazards can easily become significant, resulting in elevated risks. The majority of the work involves interaction with subsurface conditions defined during design by limited information. Experienced design engineers and owners typically involve specialized trenchless engineers early in a trenchless project to help identify hazards and mitigate the consequences of encountering hazards. These specialized engineers can provide assistance in project alignment selection, providing constructible project descriptions used for permitting, review, and input into project documentation for construction, and construction support.

5.2 PLANNING AND PERMITTING

5.2.1 The Process

Concept planning on a project entails defining the owner's objectives and tolerances needed for a successful project. This is followed by assessment of alternative construction techniques by an experienced trenchless engineer, often called *proof of concept*. The result is an owner/project engineer determination that pipe ramming is a feasible and affordable construction process. Note that the decision that pipe ramming is a constructible process for meeting the owner's requirements has been made by the owner and engineering team. Because the contractor did not select the pipe ramming process, they have a right to assume that the specified process of pipe ramming is constructible using industry standard practices as modified by specification. The following sections discuss approaches to determine the validity and feasibility of using pipe ramming for construction.

The planning and permitting process assesses the interaction of the pipe ram construction with anticipated and assumed subsurface and environmental conditions to identify and manage risks of encountering hazards on project cost, schedule, environment, and safety. The basic premise of hazard identification is that there is no hazard if there is no project. Therefore, all hazards relate to the project.

Mitigation of hazards may be accomplished by avoidance, acceptance, sharing by contract, quantification of design, and permitting conditions for constructability. Because hazards are the result of the interaction between the construction and the site-specific conditions, the engineer must understand the constructability requirements for a successful pipe ramming construction process. Interactions are considered hazards. The hazard significance is assessed by qualification, quantification, and likelihood of occurrence. Assessment involves discussions with the project owner to provide information for the owner/engineering team to initially determine cost, schedule, and importance. This may be accomplished by simply asking the owner if their project decisions will be based on either cost or schedule. There can be only one answer. Once that information is determined, the scope of studies and design can be established for anticipated hazard mitigation. An example of the process is a comparison of scope, cost, and schedule for a pipe ram installed in a remote rural area under a rural highway versus the same ram installed in an urban residential area. The rural situation may result in a very basic design scope, whereas the urban design with the same subsurface conditions may require a major design effort because more hazards are present that have a higher risk of project impact.

5.2.2 Cost and Schedule

The importance of a hazard as rated by owner and engineer assessment is based on the potential for impact to project cost and schedule. For this reason, it is important to quantify cost and schedule impacts at the beginning of a project based on discussions between the owner and engineer. This communication between the engineer and owner is one of the most important risk management tools. See Chapter 3, Sections 3.5.1 and 3.5.2, for an initial discussion of the direct and indirect costs of pipe ramming.

Base planning cost and schedule should be prepared by experienced personnel and include the major cost line items associated with pipe ramming at a specific site with conditions defined by local experience. Estimates for labor rates may depend on prevailing wages and union/owner project requirements. The line item base cost and schedule typically include

- Mobilization and demobilization as lump sum including
  - Hammer mobilization and demobilization costs,
  - Compressor costs, and
  - Crew mobilization and demobilization costs.
- Casing cost delivered to site per linear foot of casing.
- Casing fabrication as a daily rate
  - Loaded assembly crew cost,
  - Casing handling equipment like excavator and rollers, and
  - Welding or other connection equipment.

- Launching and receiving shaft lump sum
  - Preparation cost including rail alignment system with concrete foundation.
- Pipe ramming crew and equipment as a daily rate
  - Hammer base cost per monthly rental,
  - Crew loaded rate per day, and
  - Compressor daily rate and fuel costs per day.
- Cleanup as a lump sum.
- Base location mitigation items as a daily rate.
- Contingency percent as applied to project base cost items, such as
  - Fabricate pipe production rate,
  - Pipe ramming production rate, and
  - Drive length.
- Premium costs.

Contingency costs are often estimated as a percentage of project cost assessed by varying the units associated with the base cost. These units would be the estimated production rate, estimated fabrication rate, and base daily cost for contingencies such as traffic control, work hour limitations, and weather delays.

Premium project costs must be added to the base cost and base cost contingency to address site-specific hazards. These hazards may include construction variations between anticipated and found field conditions, third-party impacts, and environmental impacts.

5.2.3 Hazard Identification

Pipe ramming construction involves processes that interact with the environment. Hazard assessment and determining possible mitigation options make up what is known as a *feasibility study*. Hazard acceptance, acceptable permit conditions, and availability of cost-effective mitigation options can determine if the project is constructible at the given location. The following project-specific hazards may be associated with pipe ramming construction. Management of the risks associated with these hazards will influence whether a project is considered successful or not successful.

The following list presents possible concerns associated with pipe ramming and is organized by the different project stakeholders. The list is not intended to be all inclusive.

1. Social/regulatory authority/permitting agency concerns:
   - Site security for safety around shafts and heavy equipment;
   - Regulatory requirements governing design and construction;
   - High noise levels from hammer impact and exhaust during casing advancement;
   - Ground vibrations during driving;

- Noise and exhaust from support equipment like excavator and welding equipment;
- Groundwater discharge and permit;
- Dewatering activities including hazardous materials in groundwater;
- Traffic management;
- Damage/disruption to existing infrastructure: environment, private property, and adjacent businesses.

2. Owner/engineer concerns:
    - Obtaining project funding;
    - Right-of-way site access to shafts for construction;
    - Land acquisition of sites with a suitable work zone for launching and receiving, storage, and casing fabrication;
    - Determination of construction schedule, such as need or cost recovery;
    - Understanding of pipe ramming standard practices, including project hazards, risk mitigation options, and impacts;
    - Hazard identification;
    - Understanding of project engineering and construction tasks to estimate project contingency costs;
    - Obtaining manageable regulatory conditions and budget associated with pipe ramming industry practices for construction;
    - Working space for construction;
    - Clearance of utilities and other infrastructure to prevent damage;
    - Clear definition of who is best qualified and thus responsible for managing various risks;
    - Specification of structural materials and joint fabrication to permit a design and to have all bidders using the same materials and procedures for bid preparation;
    - Tolerance and right-of-way associated with the capabilities of a nonsteerable technique;
    - Definition of subsurface conditions and buried objects to obtain reasonable construction bids; and
    - Definition of a completed project, payment measures, and schedule.
3. Contractor concerns:
    - Joint strength sufficient for dynamic hammer loads;
    - Casing stiffness to transfer energy to leading edge without buckling;
    - Connection of hammer to casing stiffness to maximize hammer efficiency;
    - Hammer guide rail system stability and setup accuracy;
    - Stability of the ground in the shafts for guidance and laterally and above the casing crown, up to approximately three pipe diameters from the casing;

- Penetration rate;
- Spoils removal and removal rate;
- Hammer power and efficiency to overcome ground friction;
- Ground abrasion damage to casing, linings, and coatings;
- Limited to no steering ability for correcting line and grade;
- Limited ability to remove buried objects, typically buried objects up to about one-third the diameter of the casing can be removed relatively easily;
- Stability of ground supporting feature being crossed by pipe ramming under dynamic cyclic loading;
- Encountering the subsurface conditions anticipated from project documents allowing the use of pipe ramming as the means and methods;
- Definition of subsurface conditions such that the contractor can determine equipment types, sizes, production rate, and operations for constructing the project;
- Definition of allowable workspace in project documents;
- Clearances from utilities to mitigate damage from ground vibrations and ground movement that may occur because of use of reasonable pipe ramming means and methods;
- Definition of buried objects and contract means for fairly handling unanticipated conditions;
- Constructible tolerance for line, grade, and casing size to permit placement of the carrier pipe at the design line and grade; and
- Payment.

The majority of high impact hazards are related to the interaction between the pipe ramming equipment and subsurface conditions. These are discussed further in the following sections and include

- Geotechnical conditions:
  - Penetration rate,
  - Heave and/or settlement,
  - Ground movement impacts to adjacent utilities or other infrastructure,
  - Design of the cutting head,
  - Uniformity of ground with respect to deflections from design line and grade,
  - Mechanics of ground movement during the drive into the casing or away from the casing during all portions of the drive, and
  - Impact to required casing stiffness and size to allow achievement of carrier pipe line and grade.
- Groundwater level:
  - Stability of subsurface conditions at the face of the drive pipe, and

- Stability of the soil surrounding the pipe barrel during the drive.
- Structural considerations: Determine the hammer size required for the crossing.
- Casing structure: Diameter and wall thickness to withstand driving force and for the design length and production of an opening with sufficient size to install the carrier pipe on line and grade.
- Site constraints/shaft size and depth that determine casing length of each drive segment and amount of casing coupling or welding during construction.

5.2.4 High Risk Hazards

With the design of any trenchless project, the risks should be identified and addressed as discussed in this chapter and in Chapter 3. Typical risks associated with pipe ramming relate to managing the stability of the soil face at the leading edge of the drive. Insufficient subsurface characterization may result in one or more of the following problems:

- Selecting too small of a hammer size and thin casing wall,
- Hitting any buried object,
- Ramming with insufficient ground cover,
- Encountering mixed-face conditions with significant contrasts in ground stiffness/density,
- Encountering unidentified groundwater,
- Losing line and grade,
- Developing higher than anticipated friction forces,
- Damaging the shoe or steel casing,
- Creating ground settlement, and
- Damaging adjacent utilities.

5.2.5 Permitting and Design for Special Crossings

Some state and federal transportation agencies and railway owners have established requirements for trenchless crossings. These requirements often include details pertaining to minimum installation depth and length; casing requirements, including wall thickness; preferred installation methodologies; and orientation of the installation. It is important to comply with special conditions that may be required by special crossings under railroad property, on airport property, at Transportation Security Administration (TSA) secured sites, and under highways. Crossing-specific requirements may be found in various manuals produced by AREMA and individual railroads, the Federal Aviation Administration (FAA), FHWA, and various state utility accommodation manuals.

5.3 EXPLORATION

5.3.1 Planning an Exploration Program

Explorations of surface and subsurface conditions are part of the due diligence field studies necessary to support assumptions that site surface layout, subsurface utility, and subsurface conditions are suitable for pipe ramming. Ground truth of surface conditions requires site visits, surveying, and explorations to characterize subsurface conditions. Ground truth verification of constructible subsurface conditions requires physical sampling of subsurface conditions and characterization of properties. Pipe ramming is a suitable process for quickly and cost-effectively installing pipes. However, certain subsurface conditions are not suitable for the pipe ramming process or may prove too expensive to effectively mitigate, thereby making pipe ramming not feasible. Something as simple as assuming that a plan and profile are in suitable ground may negate the use of pipe ramming should the ground truth data indicate the proposed pipe ram horizon is in rock and not soil.

Significant hazards should be verified with ground truth data to assess risk to safety, cost, and schedule should the hazard be encountered. Geotechnical engineering experience has shown that explorations completed with no idea of their overall purpose often add little value to a project. Therefore, the majority of this section examines the various explorations that can contribute to the success of pipe ramming projects.

5.3.2 Site Layout

Baseline data for planning a usable site layout and access are required for a successful project. The basic purpose of the site layout is to reliably know where you are in relation to all known surface and subsurface buried objects and conditions. The accuracy and work associated with defining a site layout are relative to the location, number, and frequency of hazards. Typically, satellite images are not suitable for constructing this layout; rather a site survey by a licensed surveyor is needed to determine the following:

- Property ownership in work areas and access to work areas,
- Utility locations to a Subsurface Utility Engineering (SUE) Quality Level A for the entire survey area that will cover both the ram installation and the launching and receiving shaft construction,
- Surface details and elevations for the pipe ram corridor a minimum of 100 ft wide (30 m) along the proposed pipe ram centerline,
- Surface features and elevations of the launching and receiving workspaces within a minimum 100 ft^2 area (9.3 m^2) around the proposed launching and receiving shaft area, and

- Locations of overhead wires or other overhead obstructions to construction lifting requirements within the construction work area.

Site layout is impacted by the scope of the construction work and pipe size along with permitting conditions and right-of-way limitations for obtaining permanent and construction easements. The amount of equipment required for ramming presents a relatively small footprint compared to that required for other pipe installation techniques. Spoils can be removed from the pipe with air pressure, by sluicing, by small equipment such as skid steers and HDD rigs, or by augers; however, removal by augers is the most common method. As such, the required launch site layout area may be considered similar to the layout area at a launching shaft for an auger boring operation. The receiving shaft needs to be large enough to retrieve a multitude of components—such as cutting shoes, pilot tubes, adaptors, and crushing tools. All of the components are sufficiently expensive, and the contractor normally recovers these items at the end of the ramming process for use on a subsequent ram. Other contractors may leave the cutting shoe in place, allowing the subsequent installation of the carrier pipe to dictate the excavation at the end of the ram. Alternatively, they may jack in a secondary pipe to be the carrier, thus requiring removal of the jacking pipe by cutting the pipe into sections as it is extruded into the receiving shaft.

The area required at the launching shaft is dictated by the desired length of the casing pipe and the length of the hammer with pneumatic or hydraulic supply lines, fully assembled with the pneumatic lines that attach to the compressor or hydraulic lines that attach to a power pack. Surface equipment typically consists of a crane/boom truck/excavator to lift and set casings into place for ramming, the spoils removal system, welding equipment, and a tool trailer. It is also necessary to have a dump truck or area to store the spoils that are removed from the pipe.

Most contractors prefer 20 to 40 ft long (6.1 to 12.2 m) casings to minimize the welding time. The welding process typically controls the overall advance rate of the installation; therefore, minimizing welding can markedly decrease project duration. Fully assembled hammers can add a significant length to the shaft or area required for ramming. It is typical to require a 40 ft long shaft for pipe ramming a 20 ft long casing if the casing diameter is 36 in. (915 mm) or larger [requiring a hammer and assembly that is approximately 12 to 14 ft long (3.7 to 4.3 m)]. For smaller diameter rams, shaft lengths of approximately 30 ft (9 m) are more common because the hammer is shorter in length.

Ramming sites should provide ample room around the launching shaft to allow for shaft construction, support equipment, and casing storage (or delivery of casing), and should consider access and egress routes for trucking spoils from the site. If ground stabilization such as dewatering or lateral ground support is required, then it may be necessary to add room for both the dewatering system and excavation bracing system to have unimpeded access for the ram equipment and pipe sections.

Provide the results of the site layout survey with construction and permanent easements on the project drawings, along with coordinates and elevations at each end of the ram to verify the actual design location of the crossing. Site layout for design does not need to show the construction equipment layout. The site layout needs to show adequate work area where the contractor will be able to lay out the equipment. Excessive work area restrictions add cost to the project. Hence, a design objective is to develop a way that the project can be constructed and provide sufficient room for the assumed construction method.

5.3.3 Geotechnical Considerations

Pipe ramming is a rapid and versatile trenchless method, allowing installation in a wide range of soils—from loose flowing sands, to very dense soils with cobbles and boulders. The process is based on horsepower, and the general rule is "If it does not move, get a bigger hammer." Operator experience is necessary to operate the hammer appropriately in different ground types to avoid overexcavation, heave, or stalling the drive (insufficient horsepower). The hammer energy and frequency may be adjusted to a limited extent by adjustment of air or hydraulic fluid pressure and volume. These changes allow the operator to find the "sweet spot" for the optimal penetration rate in a specific ground type and density. This ability to adjust the hammer output energy and frequency has resulted in cases in which the casing was "steered" to a limited extent up or down by changing the energy and frequency. However, this steering is not predictable or easily controlled without some form of locating sonde on the front of the casing to know where it is, so the operator can try to adjust where he or she wants it to be.

Geotechnical conditions will be affected by the ram process, possibly resulting in surrounding ground displacement and vibration. These interactions can cause adverse reactions to adjacent structures and utilities or heave/settlement to overlying structures like pavements. Prediction of the reaction may be assessed by empirical methods and by dynamic characterization of ground and groundwater properties along the drive path. However, design choices only reduce the magnitude of the impact to a limited extent. The risk of adverse reactions remains.

Mixed-face conditions consist of two different soil materials at the face of the pipe ram. This can cause a variety of problems including deflection of the drive path and overexcavation. Overexcavation occurs when denser ground on one side of the casing slows the advancement rate and the looser ground is able to ravel or flow into the casing. Although mixed-face conditions do not always result in negative consequences during ramming, it is typical to see a change in ramming rate (ft/min), either positive or negative as the ram transitions into the softer/looser or harder material. Loose material in the pipe ram horizon can become unstable from vibrations as a result of raveling or flowing conditions. The combination of vibrations, induced energy, and groundwater could lead to liquefaction of the loose soils and a reduction of soil strength.

The geotechnical conditions dictate many of the empirical design factors used by contractors. Design parameters include the configuration and design elements of the cutting shoe, the thickness of the casing to avoid buckling, use of pilot tube for guidance, and the launch strategies.

5.3.4 Groundwater

Pipe ramming is commonly used above the groundwater. However, the method can be used below the groundwater table in a variety of ground conditions. The key to using pipe ramming below the groundwater is to maintain the excavation face stability to avoid flowing ground conditions into the casing. This is typically accomplished by installing a plug at the front of the first casing section, inside the cutting shoe that spans the inside diameter of the casing (Figure 5-1). This provides a counterbalancing weight and prevents a flowing ground condition, specifically the ground above and in front of the casing flowing into the face.

The groundwater pressure should be measured prior to pipe ramming to plan for an appropriate plug length within the casing. A completely watertight plug is not recommended because ramming forces can become extremely pressurized. The intent of using a plug at the front of a casing is to keep the ground from flowing in the casing during the pipe ram. Allowing some water to enter into the rammed casing prevents high groundwater pressure to build up at the face. This is the primary reason that sandbags are often used as plugs because they allow the groundwater to flow through while keeping the ground in place.

5.3.5 Ground Settlement or Heave

If pipe ramming is performed in loose sandy or silty soils, the soil above and below the pipeline can be densified or locally liquified by the vibrations, resulting in ground settlement and loss of ground support. Soils prone to settlement from vibrations will not continue to settle after the ramming process is completed because the vibration is only construction related. Ground settlements should be assessed before, during, and after ramming to determine if the amount of settlement can be tolerated by the project. The loss of ground support typically results in changes to line and grade.

Figure 5-1. Sandbags in the leading casing used as a plug to resist flowing ground conditions when the casing is below the groundwater table.
Source: Courtesy of Staheli Trenchless Consultants.

If pipe ramming is performed in dense soils, the soil above, in front of, and below the casing can be driven outward, resulting in ground heave. Sometimes heave can occur when the earth plug within the casing prevents additional soil from entering the casing. This situation can be mitigated by partially removing the soil from within the casing as part of the pipe ramming process. Ground heave should be assessed before, during, and after ramming to determine if the amount of heave can be tolerated by the project.

5.3.6 Deviations in Line and Grade

Line and grade tolerances are an important design consideration. Pipe ramming is a nonsteerable technique and usually has a tolerance of 1% of the drive length when ground conditions are favorable, drives are shorter, casing walls are thicker, and casings are larger in diameter. When ramming smaller diameter casings, there is a greater chance of line and grade variation. If the soil conditions are less homogeneous and contain gravels, cobbles, boulders, or other objects that could impact the line or grade, larger tolerances should be considered. If the casing needs to be installed at a more precise tolerance, then the engineer should consider the use of a pilot tube for guidance or consider a different method.

There are steps that can be taken to improve the installation alignment and grade of the casing. Care should be taken in placing the ramming equipment at the launch shaft/area. The alignment of the casing within the shaft is a critical parameter for setting line and grade of the ram. If rails or tracks are used to place the casing in a shaft/launch area, care should be taken to ensure the tracks are set to the specified alignment. Figure 5-2 shows a launch shaft for a pipe ram where a rail system was used in conjunction with a hammer cradle to ensure that the line and grade of the casing and hammer were set to the project's design requirements.

In addition, it is important that the hammer be aligned at the same orientation as the casing. Introduction of even a slight angle at the connection between the hammer and the casing can result in loss of energy transfer with line and grade deviations. The introduction of an angle at the connection of two successive casings can result in line or grade deviations or a curved installation. The curved installations can be vertical, horizontal, or compounded curve. Using a thicker walled casing will help resist physical bending of the casing during ramming, decreasing the potential for line and grade deviations. The use of high-strength, thick-walled casing with tight dimensional tolerance, combined with acute attention to setup of the ramming equipment, allows the best chance of installing a pipe ram to a set alignment and grade in the absence of buried obstructions or other difficult ground conditions.

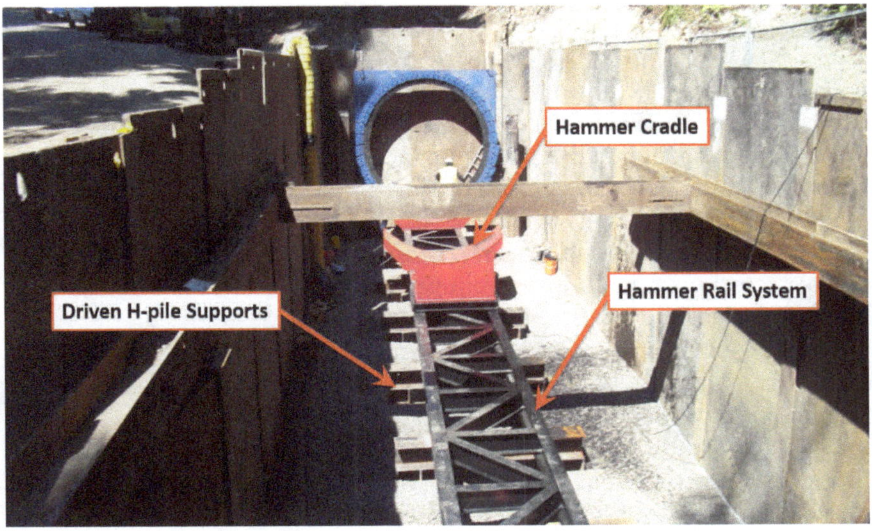

Figure 5-2. Components of a shaft illustrating rails and hammer cradles used to ensure the casing and hammer are set to the project's design line and grade. Source: Courtesy of Staheli Trenchless Consultants.

5.3.7 Appropriate Casing Size

The casing should be sized to accommodate the carrier pipe, taking into consideration any space required for casing spacers and backfill grout thickness. In situations in which the carrier pipe flows by gravity, it may be necessary to upsize the casing to allow extra space to adjust the grade of the carrier pipe within the rammed casing. It is often necessary to make adjustments in the grade of the carrier pipe because of deviations in the grade of the rammed casing during installation. If the ground through which the casing is to be rammed contains gravels, cobbles, or boulders, it is often wise to consider the tolerance of the overall ram and upsize the casing to allow the maximum deviation over the length of the drive; this will allow placement of the gravity flow pipe within the casing.

The actual annular space allowed for the casing spacers and backfill grout/material will be controlled by different factors, including

- Need for the physical working space,
- Type of casing spacers used,
- Ability to place the backfill material,
- Owner's requirements for minimal backfill thickness,
- Corrosion protection requirements, and
- Allowances for grade control of the carrier pipe.

5.3.8 Ram Length Limitations

The maximum achievable ram length is a function of the amount of driving force provided by the attached hammer versus the amount of resistance from the ground. The maximum achievable length is limited by the frictional resistance of the ground acting on the outside and inside of the pipe, weight of the spoils in the casing, and the face resistance acting on the leading edge of the shoe/casing. These resistance forces are determined by the geotechnical conditions—reduced by use of lubricants and increased by packing/bulking of ground inside of the casing.

Other contributing factors to the maximum achievable ram length include changes in geotechnical conditions, encountering denser or looser materials, encountering buried objects, introduction of angles into the casing at welds, and nonconcentric connection of the hammer. All of these factors absorb force that would otherwise be used to propel the casing forward.

Ram lengths for a specific project can be estimated based on the frictional forces, casing weight, and hammer size. Typical ram lengths based on various hammer sizes are presented in Figures 2-7 through 2-9.

5.4 DESIGN CONSIDERATIONS

5.4.1 Basis of Design

Design prior to construction is a planning process used to qualify and quantify components of the project to permit multiple parties to fairly price and schedule a project. Design assumptions and data supporting those assumptions on some projects are documented in a basis of design report. All trenchless projects should require some form of documentation demonstrating that the basis of design is founded on industry standards of professional care including assumptions, data, and analyses. Project assumptions and data used for design are identified and provided in the bid documents for use by the contractors for developing their bid. Calculations may be used in part to justify that a project is constructible.

5.4.2 Selection of Line, Grade, and Tolerances

Line and grade should be set within the industry practice tolerances without damaging surrounding structures. Depth is typically limited to a minimum of three casing diameters below the surface, or a minimum of 6 ft (1.8 m) if other utilities are present.

Precision Precision is the ability to repeatedly group a number of attempts in a target of a certain size with a specific setup. Although pipe ramming is considered a nonsteerable trenchless construction process, it typically behaves like a rifle and goes where it is pointed. Given a quality rigid setup platform, no obstructions or mixed-face conditions, and rigid support of the hammer, the system is usually capable of hitting a target in the direction aimed with a precision of +1% for drive lengths less than 100 ft (30 m) and for diameters larger than 3.3 ft (1 m). Table 7-1 provides installation tolerances from actual case histories.

Accuracy Accuracy is the ability to place a target at a specific location then aim and hit the target. Pipe ramming is nonsteerable. However, a quality alignment setup can result in an accuracy similar to the system precision of 1% of the length.

Zone of Radial Influence The surrounding ground influenced by the ramming process is typically two to three pipe diameters measured from the outside of the casing. In general, the zone of influence extends from about one pipe diameter below the casing invert up/outward to two to three diameters laterally and vertically. The zone of influence is generalized because it strongly depends on the soil type (cohesive or cohesionless), the soil density, and the soil stability (softer/looser materials are less stable than denser/stiffer materials).

Correction Ability There are some slight corrections to line and grade possible for the pipe ramming process, but usually only in the first casing rammed. The casing setup may be biased to account for rise or drop of the trajectory. Varying hammer energy and frequency can influence to a limited degree the tendency for the pipe to climb or drop along a drive.

Buried Object Deflection Buried objects can easily cause the drive line and grade to deviate an unplanned amount. Buried objects can also overload the casing structure, causing limited to extensive failure of the casing wall. This may be corrected by driving additional casing and cutting out the damaged section or by abandoning the casing and redriving, or slip lining the original casing.

5.4.3 Casing Material Selection and Dimensions

Casing selection is based on the project loads and the need for carrier pipe installation to meet specified grade lines within specified tolerances. Casing size (diameter and thickness) may be increased if carrier pipe must be set at a more precise line and grade. The determination of the final casing diameter should also account for deflection bends in the pipe and the mechanics of how the carrier pipe is to be placed on the desired line and grade: spacers, rail system, blocking along with the securing procedure to hold the pipe in place such as with grout.

Specify structural steel material that is pressure rated for casing. This material selection will ensure that the steel welds in the cans have a proper strength for the ramming process. Typically specify ASTM A139 Grade B or similar material. Grades 50 and 60 have been specified and successfully used. Use of higher-grade steel may result in the use of a thinner walled casing. However, success of a drive is the result of stiffness, not wall strength. Successful drives typically use a maximum dimension ratio (DR) of 100 and a minimum wall thickness of 0.25 in. (6.4 mm) for pipes less than 8 in. (205 mm) in diameter. DR is the pipe outside diameter divided by the wall thickness. Selection of the pipe wall thickness should be verified with hammer manufacturer charts that are based on empirical databases. Typical manufacturer recommended wall thickness to casing size has been empirically developed and is included in a variety of summary tables. Table 2-1 is a combined table for general use.

Some owners in the oil and gas industry and the transportation industry (rail and highways) may have their own design requirements for the casing thickness. These requirements would supersede any wall thicknesses calculated or obtained from the hammer manufacturers' tables or Table 2-1.

Table 2-1 notes which casing minimum thicknesses are greater than or less than values included in AREMA (2019) for E80 loading.

Minimum casing wall thickness is based on maximum loading cases. Because of the high stress conditions applied by a hammer, casing wall thickness is usually governed by drive force to depths up to 20 ft (6 m) above the crown. Deeper casings should be assessed by experienced geotechnical engineers using typical tunnel loading models that account for soil arching and the many other types of soil loading conditions based on site-specific soil properties.

5.4.4 Ground Response (Settlement, Heave, and Vibration)

The passage of a steel casing fitted with a cutting shoe through any point of soil will initially cause the ground to heave (or produce a tendency for heave) as the cutting shoe passes under the point on the ground surface of interest, and then settle as the soil compressed by the cutting shoe moves (or tends to move) into the newly created annulus behind the cutting shoe. In general, the heave and settlement of pipe ramming installations can be computed using methods established for tunneling. Commonly used empirical methods include those by Cording and Hansmire (1975), Mair and Taylor (1997), and O'Reilly and New (1982), which are based on an assumed distribution of ground loss that reaches a maximum above the center of the casing and reduces with distance transverse to the longitudinal axis of the casing. These tunneling-based methods assume that ground loss or heave is the sole source of deformations. However, in pipe ramming installations, the stress waves imparted to the pipe by the hammer will also generate ground vibrations (as discussed in Section 5.4.6). The ground vibrations contribute to settlement and increase the distance of influence relative to typical ground loss models.

However, for installations above the groundwater, settlements associated with pipe ramming installations are typically smaller than with tunneling methods, as the ground is continuously supported by the steel casing throughout the installation. In general, settlement estimates require the consideration of the effect of the overcut produced by the cutting shoe and the use or lack of use of lubrication. If lubrication is used to reduce the casing resistance and stabilize the overcut, then only part of the overcut volume should be used in the analysis of settlement (Bennett 1998). Ground arching induced by the overcut can prevent the ground from fully collapsing onto the casing. This effect can be considered in the ground loss volume calculations; Bennett and Wallin (2010) suggest that approximately half of the overcut volume may be assumed to contribute to settlement. An evaluation of settlements observed for pipe ramming installations found that the assumption that half of the overcut volume contributes to ground loss appears reasonable (Studlein and Meskele 2013); however, the breadth of settlement influence transverse to the casing was underestimated. Notably, 97% of the settlement data used to evaluate the tunneling-based settlement models was below 0.4 in. (10 mm), indicating that settlements associated with pipe ramming tend to be negligible (Stuedlein and Meskele 2013).

The process of a pipe ramming can produce ground vibrations that are largest directly above the leading edge of the casing and attenuate with distance from the casing. Ground vibrations may produce damage to nearby buried structures or adjacent buildings if the magnitude of vibration is large enough and if the frequency of vibration correlates with a fundamental modal frequency of the structure of interest. Damage can range from low impact effects, such as the onset of cosmetic cracking in architectural cladding, to structural damage that can prevent serviceability or reduce structural stability. In general, ground vibrations are assessed with Office of Surface Mining (OSM) and United States Bureau of Mines (USBM) criteria using the peak vector sum (PVS) velocity, which represents contributions of longitudinal, transverse, and vertical motions induced by the soil–pipe response to the hammer impact and the frequency content of the constituent motions. These criteria specify maximum allowable PVS velocity as a function of the frequency content, noting that the maximum permissible PVS velocity may increase with increases in the frequency of the motion.

Observations of pipe ramming–induced vibrations reported by Meskele and Stuedlein (2015) indicate that a wide range of amplitudes and frequencies is possible from pipe ramming, with maximum PVS velocity amplitudes ranging from 0.04 to 4 in./s (1 to 100 mm/s) and 20 to 100 Hz. In general, PVS velocities of up to 0.8 in./s (20 mm/s) are permissible at 20 Hz, increasing linearly to 2 in./s (50 mm/s) for frequencies greater than 40 Hz. Based on the observations of pipe ramming–induced vibrations, Meskele and Stuedlein (2015) proposed several models to predict the best-fit average and upper-bound variation in PVS velocity amplitude with distance from the leading edge of the casing as a function of energy transferred from the hammer to the casing based on the power law approach by Wiss (1981). Owing to the geometry of rammed casings, separate models were developed for spherically expanding vibrations propagating from the face or leading edge of the casing (termed *forward propagation*) and vibrations propagating cylindrically along the length of the casing (termed *lateral propagation*). Forward propagating vibrations tend to be of larger magnitude near the casing but attenuate quickly, whereas laterally propagating vibrations tend to be of smaller magnitude but decay in amplitude more slowly with distance. These models may be used to plan alignments and inform risk analyses in the planning phase of a pipe ramming installation. Figure 5-3 compares the best-fit models for forward and lateral propagation of pipe ramming–induced vibrations to other commonly used construction equipment. Based on the degree of risk anticipated, the best-fit or upper-bound attenuation models proposed by Meskele and Stuedlein (2015) may be considered by the design team.

Additional guidance on structural load combinations may be found in the German standard ATV-A 161E (DWA 1990).

5.4.5 Joint Configuration

A midweld is used to connect two or more casing joints to form a longer casing string. The weld should be properly made to ensure that tensile installation stresses can be adequately transmitted during driving without pulling apart the two portions of the casing, and to avoid leakage and infiltration while in service. Welds for pipe ramming should be full penetration butt welds to ensure that the pipe does not become damaged at the weld because of repetitive impact loading. In general, the casing thickness that is used in pipe ramming is sufficiently thick that corrosion protection coatings are not required. Corrosion prevention coatings are not used because coating may become damaged during ramming or scoured as it

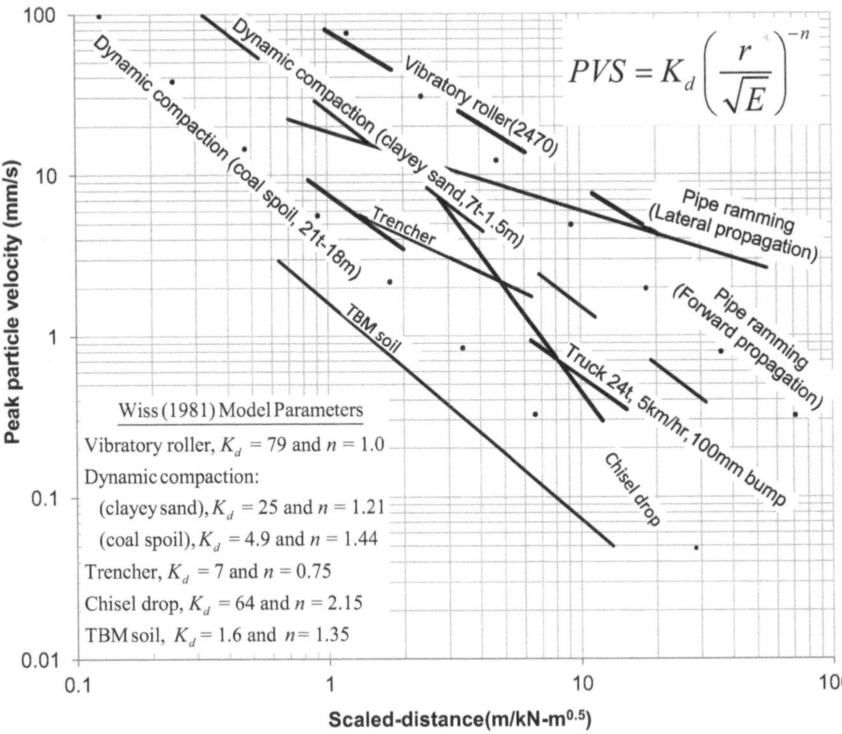

Figure 5-3. Scaled distance and peak particle velocity.
Source: Modified from Meskele and Stuedlein (2015).

passes through abrasive ground. In extremely corrosive environments, it is therefore recommended that an increased wall thickness be used to provide greater longevity. Recently, however, epoxy coatings have been used successfully for some gas transmission pipes and may be implemented for difficult pipe ramming operations. If the epoxy coating is able to withstand the abrasive effects of the installation, the coating will decrease the friction between the steel pipe and the ground, thus facilitating an easier pipe ramming operation. The epoxy coating could also provide additional protection in corrosive soil environments.

5.4.6 Driving Resistance of the Casing

Prior to advanced studies on the mechanisms that controlled pipe ramming behavior, calculation models were developed and used by some of the hammer manufacturers to determine the size of hammer that was needed for a particular project, based on generalized soil parameters, size of casing, a face resistance force, and friction coefficients. These formulas have been widely used for years, but little information is available regarding their ability to predict penetration resistance or reliability on selection of hammer size. Nevertheless, before any modeling studies or comparison of modeled behavior to actual installation data, these equations were one available method used to aid in the construction of pipe ramming. However, these equations have not been used for pipe ramming design by the engineering community in the United States. They are presented here as a way of determining how the contractor will likely select a hammer for a pipe ramming project that has been designed as described herein with models that mimic the actual behavior of the forces and how they are imparted on the casing during the ramming process.

Each manufacturer uses different equations and empirical knowledge to estimate the required force and select appropriate hammers. Typical equations to estimate the penetration resistance are based on a combination of four components: the resistance of the face, the casing friction, the weight of the ground, and the weight of the casing. The manufacturers present the following as the estimation of required force:

$$Q_R = F_f + F_c + W_s + W_p \tag{5-1}$$

where

Q_R = Estimated horizontal force (kN),
F_f = Front resistance (kN),
F_c = Casing friction (kN),
W_s = Weight of the ground (kN), and
W_p = Weight of the steel casing (kN).

Each of these are given by the equations shown subsequently. The front resistance (kN) is

$$F_f = A_R \times \sigma_s \tag{5-2}$$

where A_R is the cross-sectional area of the steel casing wall (cm^2), and σ_s (kN/cm^2) is the front resistance coefficient shown in Table 5-1.

Casing friction (kN) is

$$F_c = \Sigma U \times L \times \tau_m \tag{5-3}$$

where

ΣU = $2 \times \pi \times (R_i + R_o)$ = Internal and external circumference of the pipe (m),

R_i and R_o = Internal and external radii of the casing (m),

L = Casing length (m), and

τ_m (kN/cm^2) = Frictional force coefficient shown in Table 5-1.

The weight of the ground (kN) is

$$W_S = V \times \gamma_s \tag{5-4}$$

where $V = \pi \times R_i^2 \times L(m^3)$, and γ_s is the weight of soil coefficient (kN/m^3) shown in Table 5-1.

The weight of the steel casing (kN) is calculated by

$$W_p = L \times \gamma_p \tag{5-5}$$

where L is the length of the casing (m), and γ_p is the weight of steel casing is 493 lb/ft^3 (78.5 kN/m^3).

5.5 DOCUMENTATION

Design documentation of a pipe ramming project is typically used to clarify the project features and assumptions for retaining an independent contractor to schedule and cost a project. In general, design documentation may include

- Design feasibility memoranda;
- Design calculations and analysis;
- Permit applications;

Table 5-1. Soil Coefficients for Estimation of Required Horizontal Force.

Soil type	Front resistance coefficient (kN/cm²) = σ_s	Frictional force coefficient (kN/cm²) = τ_m	Weight of soil coefficient (kN/m³) = γ_s
Cohesive, low density	0.5	0.003	19
Cohesive, medium density	0.6	0.004	20
Cohesive, high density	0.85	0.005	21
Noncohesive, low density	0.4	0.004	17
Noncohesive, medium density	0.5	0.005	20
Noncohesive, high density	0.65	0.006	22
Marl	0.9	0.01	18–23

Source: Courtesy of TT Technologies.

- Utility notifications/utility investigations and conflict resolution; and
- Development of contract documents to define the project and success criteria consisting of
  - General conditions,
  - Special or supplementary conditions,
  - Technical specifications,
  - Plans,
  - Geotechnical investigation and reports forming the basis of the design and specified performance, and
  - Permit conditions.

Each design step represents a general level of completion of design elements typically organized by the time necessary for completion such that all design steps are completed by the time of the bid package release. The amount and technical depth of supporting project documentation for each step often will be defined by overall project economics and risks. Reduced documentation does not mean that steps are skipped, only that steps have been condensed or simplified based on site-specific conditions or possibly by economics or failure to identify and assess the risk of encountering a hazard.

REFERENCES

AREMA (American Railway Engineering and Maintenance-of-Way Association). 2019. *Guidelines for railroad grade separation projects*. Lanham, MD: AREMA. http://www.arema.org.

ASTM. 2016. *Standard specification for electric-fusion (Arc)-welded steel pipe (NPS 4 and over)*. ASTM A139. West Conshohocken, PA: ASTM.

Bennett, R. D. 1998. "Jacking forces and ground deformations associated with microtunneling," Ph.D. thesis, 487. Univ. of Illinois at Urbana-Champaign, IL.

Bennett, D., and K. Wallin. 2010. "Aerojet groundwater extraction and treatment system area 3 Wells and GET 'H' pipeline: Highway 50 crossing feasibility study." Technical Memorandum, Folsom, CA.

Cording, E. J., and W. H. Hansmire. 1975. "Displacement around soft ground tunnels." In *Proc., 5th Pan American Congress on Soil Mechanics and Foundation Engineering*, Buenos Aires, Argentina.

DWA (German Association for Water, Wastewater and Waste). 1990. "Structural calculation of driven pipes." ATV-A 161E. Bonn, Germany.

Mair, R. J., and R. N. Taylor. 1997. "Bored tunneling in the urban environment." In *Proc., 14th Int. Conf. on Soil Mechanics and Foundation Engineering*, 2353–2385. Rotterdam, Netherlands: AA Balkema.

Meskele, T., and A. W. Stuedlein. 2015. "Attenuation of pipe ramming-induced ground vibrations." *J. Pipeline Syst. Eng. Pract.* 7 (1): 04015021.

O'Reilly, M. P., and B. M. New. 1982. "Settlements above tunnels in the United Kingdom—Their magnitude and prediction." In *Proc., Tunneling*, 173–181. London: Institution of Mining and Metallurgy (IMM).

Price, C., and K. Staheli. 2013. "Examining pipe ramming design—A forensic analysis of a high risk project." *Proc., North American No-Dig 2013, Dallas, Texas*. North American Society for Trenchless Technology (NASTT).

Staheli, K., A. Stuedlein, and P. Richart. 2018. "Pipe ramming understanding the forces that drive the industry forward." *Proc., North American No-Dig 2018, Dallas Texas*. North American Society for Trenchless Technology (NASTT).

Stuedlein, A. W., and T. Meskele. 2013. "Analysis and design of pipe ramming installations," Final Rep. SPR-710. Salem, OR: Oregon DOT and FHWA.

Wiss, J. F. 1981. "Construction vibrations: State-of-the-art." *J. Geotech. Eng. Div.* 107 (2): 167–181.

CHAPTER 6
CONSTRUCTION

6.1 GENERAL

The following sections are based on the traditional design–bid–build process. It should be noted that other project delivery methods are available. Once final design is approved, project funding is in place, and the permits and easements acquired, the project owner will put the project out for bid.

6.2 BIDDING

On any given project, a general contractor oversees all elements of the work. The trenchless and shaft portion of the project may be smaller elements of a larger project. Pipe ramming installations involve specialty work, which is often performed by a subcontractor. Specialty subcontractors may also be used to perform shaft construction and dewatering. If the work is not self-performed, the general contractor needs to communicate the project requirements and coordinate with its subcontractors and suppliers to ensure project success.

During bidding, interested bidders (general contractors, subcontractors, and suppliers) should review the bid documents, which may include the following:

- General conditions;
- Bid form;

- Special or supplementary conditions;
- Technical specifications;
- Plans;
- Permits, easements, and other regulatory requirements; and
- Geotechnical information, including subsurface exploration logs.

It is recommended that the project owner schedule a pre-bid meeting with a site visit and provide an opportunity for all interested bidders to attend. This meeting provides an opportunity for bidders to evaluate the existing site conditions, including possible utility interference, traffic control challenges, site access, adjacent construction activities, and other site constraints. For projects that are considered difficult, complex, or have important constraints/requirements to be known or addressed by the bidders, the owner should consider making the pre-bid meeting and site visit mandatory. In all cases, the pre-bid meeting and site visit should require all attendees to sign in, and the completed sign-in sheets should be included in an addendum. The project owner should also require prospective bidders to submit all questions in writing and cite specific project documents that cause concern. All bidders should receive the responses to the other bidders' questions as an addendum.

A thorough understanding of the subsurface conditions allows the contractor to select the appropriate construction means and methods. The ground dictates the success or failure of the installation. Therefore, a key consideration during the bidding phase is how the contract language presents and characterizes the geotechnical information. The contractor should determine if the contract documents include a differing site condition (DSC) clause and interpret the DSC clause. These conditions greatly influence the contractor's risk and therefore greatly influence the contractor's bid price.

After reviewing the contract bid documents and referenced documents, the general contractor will prepare and submit a bid based on the estimated cost and schedule to build the project. Responsive bidders will include in their bid the cost for any subcontracted work and the required materials. Each contractor's bid will include appropriate markup for risk, overhead, and profit.

6.3 SUBMITTALS

The contractor performing the pipe ramming work will prepare and provide detailed submittals in accordance with the requirements of the specifications. The submittals may include working and shop drawings, catalog cut sheets, calculations, schedules, and written work plans. The extent of the submittals depends on the complexity and specific risks associated with the project. The timing of the submittals varies and includes preconstruction, construction, and postconstruction submittals.

6.3.1 Preconstruction Submittals

Preconstruction submittals may include the following elements:

- Qualifications for personnel, including
  - Trenchless key personnel;
  - Professional engineer experienced in shaft design and registered in the state where the project is located, if required in the contract documents;
  - Land surveyor familiar with working underground and in shafts and registered in the state where the project is located; and
  - Welders with appropriate and current certification.
- Pipe ramming and spoils removal equipment, such as
  - Method statement (general description of how the system works and operates);
  - Ramming system, including sizing of the air compressor and, if required, a generator;
  - Spoils removal system;
  - Guidance and steering system; and
  - Lubrication system, if required by the project documents or the submitted calculations.
- Construction work plan which may include the following:
  - A narrative describing the contractor's approach to the work, which often includes a sequence of planned construction, a schedule, and the geotechnical conditions expected to be encountered;
  - The work plan may also include a geotechnical instrumentation and monitoring program, as well as the contractor's plans for mitigation of potential issues such as surface settlement, heave, encountering buried objects, encountering perched groundwater, and a stuck casing; and
  - Permitted spoils disposal site.
- Casing details such as
  - Diameter, strength, wall thickness, joint design, and other factory fabrication details of the casing;
  - Internal and external cutting shoe and other fabrication details; and
  - Field welding requirements, including joint preparation and welding material and casing material compatibility.
- Site layout plan that should
  - Ensure that all aspects of the work remain within the established easements and rights-of-way;

- Provide a check of proximity of the equipment to nearby utilities (underground and overhead), roadways, or railroad tracks to ensure a safe operation; and
- Address legal and regulatory requirements for environmental impacts from construction impacts, which for pipe ramming include control of stormwater, noise, dust, and diesel exhaust.

• Lubrication system details (if required to) include the following:
  - Details on how the contractor plans to lubricate the casing internally and externally to reduce the friction forces during ramming.
  - The lubrication plan may also include water source, lubricant mix design, and the performance requirements that the mix design is anticipated to yield.
  - Many projects require the mix to incorporate NSF 60 (clean water) products. The NSF 60 approval is based on the lubricant remaining in the ground (NSF 2016). American Petroleum Institute (API) fluids tend not to be NSF 60 clean water approved, because the products are typically used on petroleum projects.
  - Lubricant mix design and how the mix will be changed during construction for the normal variations in subsurface conditions and the mix products.

• Spoils management plan that should
  - Calculate the bank cubic yards (BCY) and the volume of spoils to be transported from the project;
  - Consider any special material handling and disposal requirements, as well as documentation of the final disposition of the excavated material;
  - Provide evidence that the disposal site can legally accept the spoils, details of the spoils disposal location, and transportation methods; and
  - Provide weight tickets and disposal receipts to the owner during construction. In some cases, the project owner will include disposal of contaminated ground even if the project documents do not demonstrate the need. The owner may know that the encounter is unlikely but that a known site is located nearby.

• Estimate of ramming force and transfer of forces, which is used to confirm that the
  - Pipe ramming hammer and equipment setup are adequate,
  - Casing can withstand the axial load that will be applied during installation, and
  - Casing can withstand the forces required to enable the installation.

• Shaft design and construction that should be in accordance with OSHA requirements:

    - Provision of a professionally engineered design should be included when required. If a design is required, the design is to be signed and stamped by a qualified professional engineer (P.E.) in the state where the project is located.
    - If necessary, the launching shaft design should include provisions for a reaction block (or similar) intended to resist the loads applied by the pipe ramming or spoils excavation equipment.
    - Launching shaft design should include provisions for an invert slab to set the casing along the design line and grade and permit for full circumferential welding.
    - Shaft design should be developed to accommodate the penetrations required to launch and receive the casing.
    - Penetration should include appropriate seals to prevent groundwater inflow as required.
    - Penetration and subsequent drive should include appropriate methods to secure the face from undesirable ground behavior during ramming operations. Common undesirable ground behaviors include flowing and running ground.
- Dewatering plan:
    - If the pipe ramming alignment is at or above the groundwater table, dewatering may not be required. If the pipe ram is located below the groundwater table or if dry shafts are required, then consideration must be given to whether it is permissible to lower the water table to achieve an above-the-groundwater result. It is important that the dewatering effort reduce the groundwater levels assumed by the shaft designer.
    - If the project is completed below the groundwater table, appropriate engineering controls are required to provide a safe work environment. Some contractors use a combination of sandbags and filter fabric to stabilize the face and allow the earthen materials within the casing to drain.
    - The dewatering plan should provide a summary of dewatering methods, estimated pumping volumes, groundwater treatment, legal disposal, and an estimate of time required to lower the groundwater table to the desired elevation.
- Contingency plan:
    - Trenchless projects involve unique contingency planning requirements. In general, the contract documents require the contractor to submit contingency plans for specific events that could reasonably occur. The contingency plan explains how the contractor will react to specific events and how it will mitigate the risk associated with this event. Some of these specific events could include
        - excessive ground movement (subsidence and heave);

    - excessive ramming force;
    - undesirable ground behavior during ramming operations;
    - damage to the casing;
    - excessive deviation from the target path;
    - damage to existing utilities;
    - dewatering problems in the launching or receiving shaft; and
    - encountering buried objects.
- Safety plan:
    - All work must be developed in accordance with OSHA standards, or more stringent local/regulatory standards when applicable.
    - In some cases, depending on the diameter of the casing installation, internal access to the casing by construction personnel may be advantageous. This might require development of a site-specific confined space plan in accordance with OSHA Safety and Health Regulations for Construction (29 C.F.R. 1926, Subpart P, Excavations).
- Ground stabilization details may be required.
- Preconstruction surveys and signed right-of-entry from private property owners and occupants:
    - One method of risk management is to conduct a preconstruction survey of surface features such as sidewalks, pavement, and curb lines, as well as facades of nearby buildings and other structures, to document the existing preconstruction condition of nearby structures above, at, and below grade. This may also include the interior lower floors of nearby sensitive structures.
    - The preconstruction survey could include a video recording with narration, photographic evidence of conditions, and possibly a geodetic survey of specific features. When possible, include scales documenting the existing condition within the photograph.
    - Preconstruction surveys are intended to protect all parties from third-party claims.

6.3.2 Construction Submittals

Construction records provide details of the work performed during construction to allow the owner to determine whether the casing installation is progressing satisfactorily, and if not, whether changes to the methodology can be implemented. These records could include

- Construction progress, including documentation of any unusual occurrences or construction challenges encountered;
- Photographs and captions;

- Schedule updates;
- Survey measurements and interpretation;
- Geotechnical instrumentation readings and interpretations;
- Volume of lubrication used, properties, and when it was used;
- Geotechnical conditions encountered;
- Dewatering details;
- Ramming loads experienced;
- Records demonstrating the air compressor and generator were performing within acceptable norms;
- Damage to the end of the casing; and
- Volume of material excavated.

6.3.3 Postconstruction Submittals

Postconstruction documentation should include

- A record drawing of as-built casing installation,
- Ground movement data and summary table, and
- Postconstruction surveys.

Postconstruction surveys should

- Document the postconstruction condition of nearby structures above, at, and below grade, to determine whether any change in condition has occurred during the construction activities);
- Duplicate as near as practical the preconstruction video record, and when possible, include scales documenting the condition within the photograph; and
- Be compared with the preconstruction survey to document any changes that may have occurred during construction operations.

6.4 MEASUREMENT AND PAYMENT

Owners may request that the work be performed on a lump sum cost basis, or on a unit cost basis. A lump sum basis means the owner will pay one fixed amount for the anticipated work. The number of lump sum items can vary on a project. The owner establishes the limits of the work included under each lump sum item. The owner might request a single lump sum cost for the entire project or a series of individual lump sum costs. Lump sum prices can be further defined with partial payments paid based on a *schedule of values* associated with each lump sum.

A unit cost basis means the owner will pay the contractor on a per-unit basis, using rates established within the contract. For unit price bids, the owner establishes a bid schedule to obtain pricing for all work associated with the pipe ramming installation. This bid schedule provides the bidders the opportunity to establish their lump sums (schedule of values) or unit prices for individual work items. A typical unit-cost-basis bid schedule could include:

- Mobilization (usually lump sum, covering both mobilization and demobilization);
- Launching and receiving shafts (lump sum, per vertical foot and how measured, or per each);
- Dewatering (lump sum or per time period), including the legal disposal of effluent;
- Installation of casing (per linear foot and how measured);
- Installation of carrier pipe within casing if required (per linear foot and how measured);
- Use of pilot tube for guidance (see ASCE 2017), if required (per linear foot and how measured);
- Manhole/access installations (per each); and
- Site restoration (lump sum).

Mobilization expenses should include transportation costs of equipment, temporary facilities as required by the contract documents, and setup of the work sites. A separate bid item for mobilization should be included because the contractor tends to have many up-front, onetime costs before actually installing the casing in the ground. These front-end costs typically include permit applications required by the contract documents, preparation and submittal of technical submittals, and preconstruction and postconstruction surveys. Mobilization is normally a lump sum bid item or a percentage of the contract amount and typically covers one-time costs associated with demobilization. If, as a result of a changed condition, the contractor incurs additional mobilization costs, these costs should be paid in addition to the original mobilization bid item.

Launching and receiving shaft costs should include design of the shafts and preparation/submittal of shop drawings, as well as the subsequent construction of the shafts in accordance with the accepted design and in accordance with the contract documents. Separate bid items for launching and receiving shafts are warranted because the work could be completed by the general contractor or by the subcontractor. In addition, the length, width, and depth of the shafts are likely to vary based on use and location, so some projects include a bid item for each shaft as a lump sum. Typically, items of work required to prepare the shaft for the pipe ramming work (such as installation of the entry/exit seals, invert leveling slab, and setting of guide rails) are paid under the pipe ramming bid item.

Pipe ramming involves the installation of a steel casing. Once the casing is installed, a carrier pipe/conduit may be installed within the casing, and that work is paid as a separate bid item. A separate bid item should be considered to pay for casing on delivery because the casing is typically specially made for the contractor's equipment and delivered in advance of installation.

Casing is usually measured and paid for by the distance from shaft wall to shaft wall, whereas the carrier pipe inserted within the casing is usually measured from center of manhole to center of manhole. If a pilot tube is installed for guidance, it is considered incidental to the overall pipeline installation. If the contractor has purchased additional sections of pipe that are not installed, they are considered incidental to the other pay items and are not reimbursed separately by the owner.

The rescue of a pipe ramming project may take several different methods and will vary based on the site-specific conditions and location of failure. Common rescue methods in personnel-entry casing sizes include hand excavation of the casing and removal of the unknown object, tunneling from the receiving shaft location and removal of the unknown object, and a rescue shaft and removal of the unknown object. Common rescue methods in nonpersonnel-entry casing sizes include tunneling from the receiving shaft location and removal of the unknown object, and a rescue shaft and removal of the unknown object. Provisions should be established in the contract to cover the cost and time associated with rescue and downtime costs. Typical causes for a rescue are the removal of unknown objects, DSCs, equipment failure, or excessive deviation from the design line and grade. The requirement of a rescue shaft is rare because pipe ramming tends to be used over short distances and the availability of the space needed to construct a rescue shaft is limited. Payment for a rescue may be an actual contractor bid item that provides for payment per vertical or linear foot of excavation. Alternatively, rescue activities can be paid as time and materials under a fixed price allowance for all bidders. The amount of the allowance should reflect the project risk. For most projects, the location and the level of effort to rescue a pipe ram project are unknown. An allowance item paid by time and materials apportions risk fairly. Compensation for a rescue should be based on the evaluation of cause. Compensation should include the contractor's downtime costs and ownership/leasing costs of the appropriate equipment.

Separate bid items may be established for the installation of manholes and other access facilities in any of the shafts constructed for the trenchless work. This work is typically completed by a different subcontractor than the pipe ramming contractor.

Site restoration expenses should include repaving the road surface, sidewalks, fencing, landscaping at the shaft locations, and bringing the workspaces back to their original conditions, usually in accordance with local jurisdiction requirements.

Demobilization expenses should include removal of equipment, transportation costs of equipment off-site, decommissioning of temporary facilities, and cleaning work sites. In most cases, the cost of demobilization is covered under the mobilization bid item. When mobilization and demobilization are included in one bid item, the percentage and payment allocated to each needs to be disclosed (e.g., 60% mobilization and 40% demobilization).

If site conditions differ from those described in the contract documents and the contractor can show the different conditions adversely affected the work, the contractor is normally entitled to some form of compensation (see Section 6.10 for details).

6.5 JOBSITE LAYOUT AND EQUIPMENT SETUP

Major permits, easements, and rights-of-ways are procured in advance of construction, usually during the design phase of the project. The contractor may be required to obtain permits such as street occupancy, spoils discharge, or dewatering discharge based on how the contractor plans to execute the work.

All overhead and underground utilities that may affect construction should be located and shown on the plans by the engineer as part of the contract documents. If the contractor is required to locate utilities through specialized survey, or *soft-dig* techniques, this should be noted in the contract documents and included as a separate bid item. Prior to digging the shafts and commencing pipe ramming operations, the contractor should call the local one-call center to survey and mark all underground utilities along the project alignment. The contractor should verify the exact location and depth of the utilities via careful test pit excavations at those locations where there are apparent utility conflicts with the project work. Test pits at shaft locations should encompass the full perimeter of the shaft footprint. Manual excavation may be required for the last few inches from the existing utilities to avoid damaging the utility. Vacuum excavation is an excellent option to expose utilities with little surface disturbance and minimal risk to the existing utility.

Temporary or permanent relocation of those utilities that will affect construction and cannot be protected or supported in place within any shaft or pipeline alignment should be coordinated in advance with the utility owner.

Sufficient space is needed around the launching shaft for the following:

- Initial space for shaft construction equipment and materials;
- Pilot tube equipment for guidance, if used;
- Lubrication mixing and pumping system, if used;
- Power supply generators;
- Ramming hammer and attachment system;
- Storage and layout of the casing;

- Additional equipment required to clean out the installed casing;
- Support equipment such as backhoe, loader, crane, and welder;
- Temporary stockpiling, loading, and hauling of excavated material; and
- Additional space for typical construction operations, including dewatering.

Section 3.1.4 provides recommended workspace requirements. The workspace includes both the space available at grade as well as the space inside the shaft. The workspace at grade should be of sufficient size for the construction of the shafts, the safe operation of all equipment, and the storing of materials necessary to complete the installation. The shafts should be of sufficient size for the free and safe movement of personnel and equipment during the setup of the equipment, as well as during the execution of all the steps of installation including the movement of materials and equipment into the shaft, ramming of the casing, and the removal of spoils from the casing and shaft.

The layout of the equipment and materials around the shaft should be planned for the efficient and safe operation of the equipment and the movement of the material with minimum handling.

The workspace should be limited to the easements and/or rights-of-way unless prior written consent is obtained from an adjacent property owner, usually after proper compensation.

Local authorities (e.g., emergency services, transit, and other agencies) should be notified of upcoming work activities in advance of mobilization to the work site. Pedestrian, vehicular, and emergency traffic considerations; noise restrictions; and the needs of the local residents are just a few of the items that may require attention. See Section 3.2.6 for further discussion of special considerations to be identified and addressed.

Figure 6-1 is an example of a typical launch shaft layout and shows an isolated shaft with no pedestrian, vehicular, nor emergency traffic control measures.

6.6 SURVEY

Prior to the start of work, the contractor will require access to established project survey benchmarks. In addition, it is helpful to provide the contractor with electronic copies of the project survey files. These will help ensure the ramming installation is completed at the correct location and at the required project line and grade. The project owner is to be notified of any discrepancies as quickly as possible to mitigate impacts to cost and schedule. Survey discrepancies, if not addressed in a timely manner, may affect the operation of the final improvement.

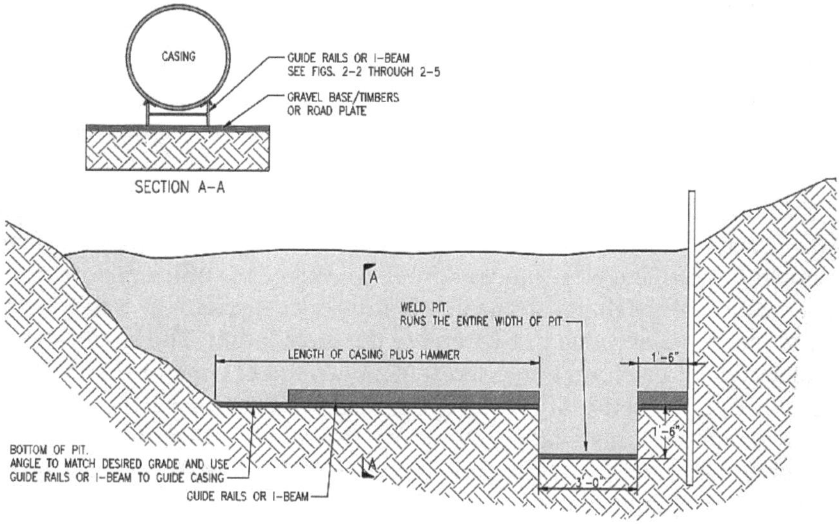

Figure 6-1. Typical launching shaft before pipe insertion and equipment placement.

Vertical accuracy relies on having properly setup guide rails fastened to the shaft invert at a specific elevation and slope. Once launched, the grade is subject to deflection, gravity, and any bias introduced through welding and casing end squareness.

Horizontal accuracy depends on precise survey marks placed at the front and rear of the launching shaft. The horizontal alignment of the guide rails is established using plumb bobs suspended from a string pulled taut between the survey marks. Once launched, the line is subject to deflection and any bias introduced through welding and casing end squareness.

A mistake in elevation at the launching shaft will be carried along the entire length of the drive. However, a small error in horizontal layout at the launching shaft will be multiplied many times over the length of the drive. Accordingly, pipe ramming requires accurate site surface survey information to be successful. The location information used by the operator is only as good as the layout information provided by the surveyor and the care with which the crew transfers that information into the launching shaft.

6.7 LAUNCHING AND RECEIVING SHAFTS

As described in previous chapters, a pipe ramming operation requires a starting location and an ending location. Shafts are required if the starting or ending location is located below existing ground surface. The shafts should be located at a safe distance from existing structures and utilities to avoid property damage and reduce risk of injury.

The locations of the launching and receiving shafts should be carefully selected, usually by the design engineer. Ideally, shafts should be at manholes and lateral connections in sewer lines, at fire hydrants and gate valves in water applications, and at splice points for high-voltage cables. Also, the distance between the launching shaft and receiving shaft should be minimized whenever possible so that friction forces do not prevent the successful installation of the pipe rammed casings or exceed the capacity of the casing.

In most cases, selection of shaft support will be at the discretion of the contractor, subject to performance criteria established in the project specifications. The shafts may need to have an engineered earth support system if space is limited, if required by easement or permit conditions, or if the shafts are deep [in general, more than 20 ft (6.1 m) in depth]. In such cases, the engineered support system should be specific to the site, prepared by an experienced engineer, and submitted for review by the designer. However, if sufficient room is available, sloping of the shaft walls may be allowed. Regardless of whether the shaft has an earth support system, all elements of shaft design and construction must comply with OSHA standards. Available shoring methods may include trench boxes and steel frames and plates. Engineered earth support systems may consist of solder pile and lagging, preengineered shoring methods, jet grouting and other soil modification methods, caissons, and steel sheet piles.

In some cases, it is feasible to design the final maintenance access points (e.g., manholes) to serve as the launching or receiving shaft. See Chapter 5 for further discussion on the use of temporary and permanent shafts.

The launching shaft is sized by the contractor and should be long enough to accommodate the pipe ramming hammer, attachments to the casing, and the length of the casing. It also should be wide enough to accommodate the workers who add casing segments and remove spoils.

After the underground and overhead interference are removed or otherwise protected and the location of the installation is confirmed with the site survey, shaft construction can begin. The project geotechnical expert should be on site to observe the excavation as it advances through the pipe ram envelope to confirm the ground conditions are as expected, as noted in Section 6.10. Once the shaft excavation has been completed, a stable subgrade and working surface is needed in the shaft bottom. This stable surface can consist of using crushed stone, gravel, steel plates, wood beams, concrete floor, or a combination as geotechnical considerations warrant for the project. The floor of the shaft is set to the proposed slope of the pipe ram.

The effect of the groundwater elevation on shaft design and construction should be considered. Dewatering efforts may be required to lower the groundwater at the shafts as well as along the alignment. In some cases, the project dewatering is limited to shaft locations only because use of the surface along the alignment is not available. (Chapter 4 provides further discussion on groundwater.) The contractor must be prepared to accommodate the dewatering equipment during pipe ramming operations and shaft construction. As is the case with every dewatering system, the contractor should take measures to prevent migration of ground. Treatment and discharge plans for dewatering operations should be developed that comply with regulatory agency requirements.

Surface water should be managed so that the shaft subgrade remains dry and stable. This can be accomplished using ditches and/or berms to divert the overland flow around the shaft. A sump pump may be needed to handle rainwater in the shaft or incidental groundwater.

6.8 STEEL CASING, LEADING EDGE TREATMENTS, AND WELDING

The contractor is required to procure steel casing as accepted in the submittal. When placing the order, the contractor should notify the supplier of the casing's intended use; end preparation for welding; and the design tolerances for roundness, square ends, and welding. The steel casing has to be properly aligned before welding to not introduce a directional bias and misdirect the casing during installation.

Leading edge treatments may include an outer shoe and an inner shoe. A typical outer shoe design has the shoe projected about 1 in. (25 mm) ahead of the casing leading edge and normally extended around the top 270 degrees, leaving the bottom 90 degrees without modification. The steel band normally has at least 0.5 in. (13 mm) of wall thickness and 6 in. (150 mm) of width.

The inner shoe is approximately the same size and extends the entire 360 degrees. It is approximately 1 in. (25 mm) behind the leading edge.

Both the inner and outer shoes require secure welding. If the ground is particularly difficult, containing cobbles and boulders, the contractor should consider plug welding the shoe to prevent separation.

The field welding, when required, must be high-quality full penetration, filled without voids, and be a proper cap of compatible welding materials preventing the rebounding tensile wave from breaking.

Mechanically joined systems like Permalok or Tri-Lock should be limited to short drives of low risk. Mechanically joined pipes tend to be less efficient in transferring the tensile wave.

6.9 SAFETY ISSUES

It is the responsibility of the contractor to maintain a safe working environment in and around the shaft and to comply with the requirements of 29 C.F.R. 1926 and local and state authority having jurisdiction. There are specific requirements for shaft construction including personnel protection using surface barricades, traffic management to protect workers within the work zone, shaft ingress and egress, and ventilation in the shaft and casing for personnel. The standard safety procedures adhered to in tunnel construction should also be followed on pipe ramming projects.

Manufacturers' safety and operations manuals should also be consulted for specific safety requirements unique to the equipment provided. Workers should understand the components of the pipe ramming equipment with special attention to the moving parts in the system, pressure hoses, and connections. It is critical that all personnel be trained to safely perform the pipe ramming operations. Workers should also be trained in the proper use of personal protective equipment including ear protection and welder safety.

For larger diameter casings, internal access of the casing by construction personnel may be necessary to facilitate installation. If so, a determination should be made as to whether a confined space exists. If so, all personnel accessing the casing should be suitably trained for confined space work. Further, adequate lighting and ventilation should be provided.

6.10 DIFFERING SITE CONDITIONS

The extent to which the subsurface conditions, including geotechnical conditions and potential buried objects (see Section 3.2.4), are identified prior to construction depends on how comprehensive the site investigation is during the design phase of the project. If identified in advance, a plan to overcome the challenges associated with the geotechnical conditions or buried objects can be developed during the design and construction phases. Unfortunately, not all geotechnical conditions or buried objects are identified before construction. Unforeseen conditions might force the casing to deviate from its intended alignment or stop the pipe ramming process altogether.

The contract documents should allow for adjustments in the project duration and price if unforeseen conditions are encountered. The contract document should require the contractor to stop and notify the project owner immediately on identifying a potential differing site condition. The owner should be notified so its representative(s) can be present to investigate and witness the situation. It is typical that the contractor does not proceed with any extra work associated with the differing condition until instructed to do so by the owner, unless an emergency situation exists. The contractor should be required to explain how the condition encountered differed from those documented in the contract. The contractor should document the time and cost impacts to the pipe ramming operation.

6.11 TRAFFIC CONTROL, FENCING, AND BARRICADING

Project locations may range from urban to rural in nature. Although specific needs will vary depending on the location, all projects require consideration of how the work will affect the adjacent community. Specific consideration must be given to pedestrian and vehicular traffic control.

During the design phase, traffic control or management of traffic plans may be developed and included in the bid documents. In these cases, the contractor implements the established traffic control plan in close coordination with the appropriate agencies. However, sometimes the contractor is fully responsible for traffic control and must develop its own plans in accordance with the needs of the local authorities, including the possible acquisition of permits to close travel lanes or sidewalks or to detour traffic. In either case, the plans typically follow the recommendations set forward in the *Manual on Uniform Traffic Control Devices* (FHWA 2012).

In addition to the traffic control around the work zone, it is important to protect pedestrians from jobsite hazards. Protective measures typically include signage, barricades, fencing, and temporary pedestrian walkways.

The pipe ramming operation will be loud. Shaft construction methods also can be quite loud. Therefore, use of sound barriers to block construction noise impacts to nearby sensitive receptors may be necessary. Typical receptors may include hospitals, schools, and residential structures. The project requirements need to be disclosed. In some cases, the noise may be limited to 5 dBA above background. This limit is nearly impossible to achieve with pipe ramming and in fact may prevent pipe ramming from being used. A more common approach is to limit work hours.

6.12 QUALITY CONTROL

Quality control (QC) is an important part of the construction phase of the project. Controls that will help ensure the casing is installed to acceptable line and grade include the following:

- Require acceptance of submittals prior to start of work;
- Verify conformance with the accepted submittals;
- Use an experienced land surveyor to provide accurate construction layout;

- Use equipment manufactured by a firm with a proven track record and that conforms to the specified performance requirements;
- Require the pipe ramming work to be performed by a qualified contractor as specified in the contract documents (see Chapter 5 for more information); and
- Verify line and grade of the casing before proceeding with the ramming process, verify again after a pipe segment is added to the pipe string, and consider using a pilot tube for guidance of the casing.

6.13 RAMMING FORCES AND LUBRICANTS

The axial load exerted on the casing is the friction between the ground and the outside surface of the casing being installed. A calculation of the anticipated ramming forces should be performed prior to the start of the work to verify sufficient casing strength is available to complete the work. At no time should the strength of the casing itself be exceeded.

There are four techniques to reduce the required ramming force. The first is to create an overcut with a cutting shoe. The second is to introduce a lubricant to the outside surface of the casing. The third is the removal of earthen material that is accumulating within the casing. The fourth is adding lubrication to the inside of the casing to reduce the friction of the earthen material moving along the inside of the casing. In stable ground, an overcut reduces the friction between the outside of the casing and the ground; a lubricant may or may not be needed. In less stable ground, the overcut will not remain open on its own; therefore, a lubricant should be introduced into the overcut to fill it, stabilize the ground, and reduce the friction.

Lubricants in pipe ramming are used to reduce friction between the surrounding and internally contained ground and the casing. The selection of lubricants depends on the type of ground through which the casing will be driven. The lubricant supplier should be consulted for guidance in determining the best lubricant to use for the project-specific geology, the proper method to mix the lubricant, and the most appropriate concentrations to achieve the desired results. Figure 6-2 shows a typical lubrication mixing and pumping system used for pipe ramming.

During the pipe ramming installation, the cutting shoe is larger than the OD of the casing, and as it advances it creates space (overcut) between the casing and the ground. In easily displaced ground, this may be sufficient to allow the pipe ram to be installed for the full length of the drive without generating excessive installation loads.

If the ground is not easily displaced, if it is overconsolidated, or if sticky clays are present, lubrication may be required to reduce the ramming forces

Figure 6-2. Typical lubrication mixing and pumping system.
Source: Courtesy of Akkerman.

required to advance the casing. The method for lubricant introduction varies by contractor.

6.14 SPOIL REMOVAL AND DISPOSAL

The pipe rammed casing will accumulate earthen materials inside. The spoils are generated as the materials are excavated or conveyed to the launch shaft. The most common method for removing spoils is the use of a horizontal auger boring (HAB) system. The HAB system's auger does not need to fill the casing. The consistency of the spoils can range from dry soil to soupy mud (especially if the spoils have been lubricated). Usually, the spoils are removed from the shaft with a skip pan (dirt bucket) or clamshell bucket lifted to the surface with a crane or an excavator. A less common alternative is using a vacuum system.

It is important to ensure that the area around each shaft has sufficient space for spoils material stockpiling in addition to the safe operation of the support equipment. Often the excavated spoils are stockpiled in the staging area around the shaft for later hauling away and disposal. If the spoils are wet, containment may be required to avoid runoff onto adjacent properties. If contaminated soil is excavated, the owner should be notified, and the contractor should take the necessary measures for safe handling, storage, and disposal in conformance with the requirements of applicable regulatory bodies. Encountering unanticipated contaminated earthen materials may require that the project be stopped until the contractor's personnel can receive appropriate training.

6.15 INSPECTION AND MONITORING

Implementation of the QC measures, outlined in Section 6.12, will ensure that the pipe ramming process is executed properly and will provide the highest likelihood of success.

Depending on the ground conditions, length of drive, diameter of the casing, and proximity to adjacent structures, the engineer may require and specify implementation of a geotechnical monitoring program to ensure that the construction does not adversely affect adjacent structures or surrounding ground. These efforts may include

- Structural monitoring points (SMPs) installed to monitor for movement of nearby structures,
- Deformation monitoring points (DMPs) to monitor for ground movement (heave or settlement) along the pipe ram alignment,
- Utility monitoring points (UMPs) to monitor buried utilities for movement during construction,
- Piezometers or observation wells to monitor the groundwater table, and
- Inclinometers installed near the launching and receiving shafts or near embankments to monitor for lateral ground movement.

Depending on the specific project conditions, the engineer may require a preconstruction survey to determine the existing conditions within the work zone. A preconstruction survey of nearby buildings and structures should document all existing cracks, cosmetic problems, and structural deficiencies. Further, elevations of SMPs, DMPs, UMPs, and observation wells should be surveyed, and the inclinometers should be calibrated prior to shaft excavation to determine baseline readings of all instruments.

In addition to establishing the required individual monitoring devices and their locations, the monitoring plan should detail the surveying method to be used to take the readings as well as the frequency and precision of monitoring and reporting protocol for the collected data. Along with the actual measurements taken, reports should include whether movement has occurred, and if so, how much. The inspectors and the contractor can then take appropriate action to mitigate any damage. A monitoring program may also establish movement limits for the various monitoring points. Reaching these limits will trigger certain actions on the part of the construction team.

6.16 REPORTS AND RECORDS

Pipe ramming equipment typically does not include automatic data logging devices, so it is recommended that the ramming activities be recorded manually on paper or an electronic device such as an iPad, tablet, or laptop. The specifications should establish the frequency with which the selected operational functions are recorded in terms of elapsed minutes, fixed distance driven, or number of times per pipe joint. The record should also include reference information, such as project name, operator's name, time, date, and casing number. The operator should also note penetration rate of the casing into the ground, number of blows per foot, air pressure on the compressor, and breakdowns and work stoppages (along with the reason for the breakdown or stoppage), in addition to observations of settlement, unusual events, lubrication added, and guidance system adjustments. A sample form for recording this information is provided in Figure 6-3.

Pipe Ramming Shift Report

Report No. ______ Date: ______ Day: ______

Contractor: ______ Shift: ______ From: ______ To: ______

Location: ______ Job No. ______ Hammer: ______

Owner: ______ Casing ID: ______ OD: ______

Face Station: Begin: ______ End: ______ Rammed Length: ______

Total Time Casing No. Minutes	List of Activities (Setting, Welding, Attaching, Ramming, Excavating)	Equipment Used	Remarks	Personnel On-Site

Observations of Settlement or Heave: ______

Comments & Instructions for the next shift: ______

Report Prepared by: ______ Signature: ______

Figure 6-3. Sample daily report form.

Section 6.3 provides a list of other information and data that could be included in a daily report as part of the construction submittals.

6.17 AS-BUILT DRAWINGS AND DOCUMENTATION

In general, as-built drawings are required for any underground utility construction, including pipe ramming projects. The contractor should mark the limits of the casing, new carrier pipe, manhole, and ancillary structure information on a copy of the contract plans. This process is known as *redline drawings*. The contractor should document any changes (vertical and horizontal) to the original layout of the underground utilities and structures that took place during the construction phase, such as rerouting a utility because of interference with the excavation of the shafts or changing the depth or alignment of the pipe to avoid difficult ground conditions. The redline drawings are usually submitted to the engineer for preparation of formal as-built documents for the owner of the facility.

REFERENCES

ASCE. 2017. *Pilot tube and other guided boring methods*, MOP 133. Reston, VA: ASCE.

FHWA (Federal Highway Administration). 2012. *Manual on Uniform Traffic Control Devices (MUTCD)*. http://mutcd.fhwa.dot.gov/.

NSF (National Sanitation Foundation) International. 2016. *Drinking water treatment chemicals—Health effects*, NSF/ANSI 60. Ann Arbor, MI: NSF.

CHAPTER 7
SUMMARY OF RECENT CASE HISTORIES

7.1 GENERAL

This chapter addresses case histories of past pipe ramming projects. Published case histories often only showcase the longest, largest, most complex, and most challenging projects. Although this section includes showcase projects, the authors have also consciously selected more common lengths and diameters. As shown on the bell curve (Figure 7-1), lengths and diameters most commonly constructed are often those least likely to be included in published case histories.

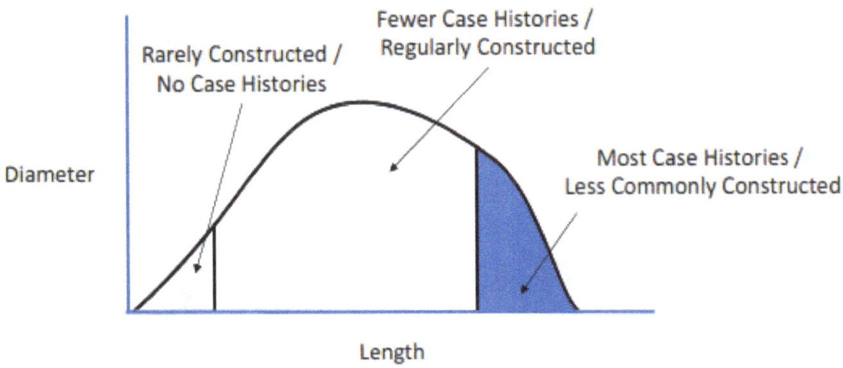

Figure 7-1. Trend of published pipe ramming case histories based on diameter and drive length.

Table 7-1 has been compiled to include both common and less-common pipe ram projects involving utility crossings, culvert installations, conductor casing installation, horizontal directional drilling assists, and pipe extractions. The 261 projects are sorted first by diameter and then by drive length. The table provides information on when the project was completed, purpose, location, casing diameter, drive length, length of the casing sections, wall thickness, hammer size, ground type, pipe ram position relative to the groundwater table, and available information on grade, as well as additional information.

Table 7.1 Pipe Ramming Case Histories

No.	Project Name	Year	Purpose	Location	Dia., in. (mm)		Drive Length, ft (m)		Casing Length/Drive	Wall Thickness in (mm)	Hammer Type	Ground Type	Pipe Ram Position to Groundwater	Design Grade	Reference
1	Exxon Mobile Gas Line Extraction	2015	Extraction of existing gas line under the Colorado River	Colorado	8	-205	150	-46			24-inch pneumatic		Below	Extraction	2016 Paper WM-T1-01
2	Puget Sound Energy Gas Mains		Cross under railroad for gas main	Seattle, Washington	10	-254									MOP First Edition
3	Puget Sound Energy Gas Mains		Cross under railroad for gas main	Snoqualmie, Washington	12	-305	120	-37	20-foot		8.5-inch pneumatic	Sand	Below w/ dewatering	2% downgrade	MOP First Edition
4	Ohio Gas Pipeline Replacement Project	2016	Extract 2,400 feet of 12-inch existing gas main	Ohio	12	-305				0.5 (13)	24-inch pneumatic			Extraction	2017 Paper WM-T1-03
5	University of New Hampshire Radiant Heating Upgrade	2012	Install twin casing through building foundation	Durham, New Hampshire	14	-355	20	-6	10-foot			Soils	Above		2013 Paper WM-T4-01
6	Sunoco Pipeline Crossing of Central Avenue	2015	Install conductor casing for HDD		14	-355	108	-33		0.50 (13)				Inclined for HDD exit	Michels case history
7	Highway 120 Crossing	2017	Install conductor casing for HDD	Dallas, Georgia	14	-355	113	-34		0.50 (13)				Inclined for HDD entry or exit	Michels case history
8	Spectra Energy Transmission Pipeline	2015	Install conductor casing for HDD	Salem, Massachusetts	14	-355	124	-38						Inclined for HDD entry	Michels case history
9	NIPSCO Crossing of the North Saint Mary River	2016	Install conductor casing for HDD	Fort Wayne, Indiana	14	-355	165	-50		0.75 (19)				Inclined for HDD entry or exit	Michels case history
10	Inergy Crossing of the Susquehanna River	2012	Install conductor casing for HDD	Wyalusing, Pennsylvania	14	-355	202	-62		0.38 (10)				Inclined for HDD entry or exit	Michels case history
11	Inter Pipeline Limited Crossing of the Vermillion River	2013	Install conductor casing for HDD	Vermillion, Alberta	14	-355	300	-91						Inclined for HDD entry or exit	Michels case history

No.	Project Name	Year	Purpose	Location	Dia., in. (mm)		Drive Length, ft (m)		Casing Length/ Drive	Wall Thickness in (mm)	Hammer Type	Ground Type	Pipe Ram Position to Ground-water	Design Grade	Reference
12	Chervon Crossing of the Mississippi River	2009	Install conductor casing for HDD	Empire, Louisiana	14	-355	500	-152		0.50 (13)				Inclined for HDD entry or exit	Michels case history
13	Fortis BC Crossing of the Kootenay River	2012	Install conductor casing for HDD	Castlegear, Alberta	14	-355	530	-162		0.50 (13)				Inclined for HDD entry or exit	Michels case history
14	Enbridge Crossing of the Kishwaukee River	2008	Install conductor casing for HDD	Belvidere, Illinois	14	-355	560	-171						Inclined for HDD entry or exit	Michels case history
15	Willamette River Crossing	2014	Install telescoping conductor casing for HDD	Gladstone, Oregon	14	-355	670	-204	20-foot	0.5 (13)	24-inch pneumatic	Gravels and cobbles	Below	9 degree entry for HDD	STC project files
16	West Palm Beach New Sanitary Sewer Porject	2001	Assist 1,000-foot (305 m) long HDD installation	West Palm Beach, Florida	18	-460					18-inch pneumatic	Sand	Below	HDD assist	TT case history
17	Cutler Site Navy Crossing	2016	Install conductor casing for HDD	Culter, Maine	20	-510	30	-9		0.50 (13)				Inclined for HDD entry or exit	Michels case history
18	Colonial Pipeline Crossing of the Nansemond River	2017	Install conductor casing for HDD	Suffolk, Virginia	20	-510	89	-27						Inclined for HDD entry	Michels case history
19	Colonial Pipeline Crossing of the Nansemond River	2017	Install conductor casing for HDD	Suffolk, Virginia	20	-510	98	-30						Inclined for HDD entry or exit	Michels case history
20	Howard Cove Navy Crossing	2016	Install conductor casing for HDD	Machias-port, Maine	20	-510	105	-32		0.50 (13)				Inclined for HDD entry or exit	Michels case history
21	Transcanada Pipeline Crossing of Beaver Lake	2015	Install conductor casing for HDD		20	-510	129	-39						Inclined for HDD entry or exit	Michels case history
22	Sprague-Roser Crossing of the Saline Creek	2013	Install conductor casing for HDD	Fort McMurray, Alberta	20	-510	131	-40		0.50 (13)				Inclined for HDD entry or exit	Michels case history

No.	Project Name	Year	Purpose	Location	Dia., in. (mm)		Drive Length, ft (m)		Casing Length/ Drive	Wall Thickness in (mm)	Hammer Type	Ground Type	Pipe Ram Position to Ground-water	Design Grade	Reference
23	Magellan Midstream Partner Crossing of Holland Bottoms Wetlands	2015	Install conductor casing for HDD	Jackson-ville, Arkansas	20	-510	195	-59						Inclined for HDD exit	Michels case history
24	TRAX Rail Line Crossing	2006	Install new water main under rail line	Salt Lake City, Utah	20	-510	215	-66	20-foot		12-inch pneumatic		Above	Pilot tube guided	TT case history
25	Highway 1 Crossing for Fortis BC	2017	Install conductor casing for HDD	Vancouver, British Columbia	20	-510	225	-69		0.75 (19)				Inclined for HDD entry or exit	Michels case history
26	Center Point Energy Crossing of the Arkansas River	2014	Install conductor casing for HDD	Little Rock, Arkansas	20	-510	328	-100		0.50 (13)				Inclined for HDD entry or exit	Michels case history
27	Duke Energy Crossing of Beverly Harbor	2003	Install conductor casing for HDD	Salem, Massachu-setts	20	-510	440	-134		0.81 (21)				Inclined for HDD entry or exit	Michels case history
28	Duke Energy Crossing of Georges Island	2003	Install conductor casing for HDD	Salem, Massachu-setts	20	-510	600	-183		0.81 (21)				Inclined for HDD entry or exit	Michels case history
29	Sunoco Pipeline Crossing of Central Avenue	2015	Install conductor casing for HDD		24	-610	37	-11		0.50 (13)				Inclined for HDD entry	Michels case history
30	Enbridge Crossing of Elleshmere Road	2008	Install conductor casing for HDD	Toronto, Ontario	24	-610	92	-28						Inclined for HDD entry or exit	Michels case history
31	Terasen-Kinder Morgan Crossing of the Clearwater River	2007	Install conductor casing for HDD	Fort McCay, Alberta	24	-610	92	-28						Inclined for HDD exit	Michels case history
32	Brunei CWDP-1 Drilling Platform	2006	Install vertical piles for launch of oil platform	South China Sea	24	-610	103	-31.5	10-foot	1.0 (25)	18-inch pneumatic		Below	Vertical	TT case history
33	Surmont Phase 1 SAGD Project	2010	Install casing sleeve under Highway 881	Fort McMurray, Alberta	24	-610	120	-37			24-inch pneumatic	Hard clay	Above	Pilot tube guided	TT case history
34	Apache Junction New Sewer Lines	1995	Install casing under Highway 88	Apache Junction, Arizona	24	-610	120	-37	40-foot		12-inch pneumatic	Caliche	Above	0.1% design grade	Hammer-head case history

No.	Project Name	Year	Purpose	Location	Dia., in. (mm)	Drive Length, ft (m)	Casing Length/ Drive	Wall Thickness in (mm)	Hammer Type	Ground Type	Pipe Ram Position to Ground-water	Design Grade	Reference
35	Northern Gas Pipeline	2005	Install 6 culverts under Beijing-Guangzhou Railway	Chansha, China	24 -610	138 -42	20-foot	0.6 (15)	16-inch pneumatic	Soft ground	Above		Hammer-head case history
36	Apache Junction New Sewer Lines	1995	Install casing under Highway 88	Apache Junction, Arizona	24 -610	170 -52	40-foot		12-inch pneumatic	Caliche	Above	0.1% design grade	Hammer-head case history
37	Texas Eastern Crossing of the Hudson River	2012	Install conductor casing for HDD	Newport Village, New York	24 -610	181 -55						Inclined for HDD exit	Michels case history
38	Caracas Storn Canal	2002	Install pipe canopy under highway and runway	Caracas, Venezuela	24 -610	200 -61	40-foot	0.75 (19)			Above		Hammer-head case history
39	Union Pacific Yard Crossings	2006	Install 3 casings under rail yard	Ogden, Utah	24 -610	200 -61	20-foot		18-inch pneumatic		Above	Pilot tube guided	TT case history
40	Niagara Gas Crossing of the Saint Lawrence River	2011	Install conductor casing for HDD	Cornwall, Ontario	24 -610	232 -71		1.00 (25)				Inclined for HDD entry or exit	Michels case history
41	Niagara Gas Crossing of the Saint Lawrence River	2011	Install conductor casing for HDD	Cornwall, Ontario	24 -610	247 -75		1.00 (25)				Inclined for HDD entry or exit	Michels case history
42	Cable Crossing of the Columbia River		Install conductor casing for HDD	Vancouver, Washington	24 -610	325 -99	40-foot		18-inch pneumatic	Soils	Below	16.5 degree entry for HDD	2015 Paper TM1-T2-02
43	Fortis BC Crossing of the Kootenay River	2012	Install conductor casing for HDD	Castlegear, Alberta	24 -610	331 -101		0.50 (13)				Inclined for HDD entry or exit	Michels case history
44	Burlington Northern Railroad Storm Water Improvements	2008	Install culvert under railroad	Rockville, Minnesota	30 -760	80 -24	20-foot		24-inch pneumatic	loose sand and cobbles	Above		Hammer-head case history
45	Wisconsin Sewer Project		Cross under Interstate 41	Wright-stown, Wisconsin	30 -760	90 -27			20-inch pneumatic	Flowing sand and cobbles	Below		2019 Paper MA-T4-04

No.	Project Name	Year	Purpose	Location	Dia., in. (mm)	Drive Length, ft (m)	Casing Length/ Drive	Wall Thickness in (mm)	Hammer Type	Ground Type	Pipe Ram Position to Ground-water	Design Grade	Reference
46	Voight's Creek Hatchery Fish Ladder	2014	Cross under Highway 162	Orting, Washington	30 -760	120 -37	20-foot	0.625 (16)		Silty sand and sandy silt	Above		2015 Paper TA-T4-01
47	Energy Transfer Company Pipeline	2018	Install conductor casing for HDD	Morgan-town, Pennsylva-nia	30 -760	138 -42					Below	Inclined for HDD entry or exit	Michels case history
48	New Subdivision in Grand Prairie	2008	Install new sanitary sewer under road	Grand Prairie, Alberta	30 -760	144 -44	40-foot	0.38 (10)	23-inch pneumatic	Hard clay	Above	0.12% design grade	Hammer-head case history
49	Cable Crossing of the Columbia River		Install conductor casing for HDD	Vancouver, Washington	30 -760	156 -48	20-foot		18-inch pneumatic	Soils	Below	16.5 degree entry for HDD	2015 Paper TM1-T2-02
50	Saratoga Road Culvert Replacement	2008	Install new culvert to replace 18-inch culvert under road	Whidbey, Washington	30 -760	170 -52	20-foot	0.5 (13)	18-inch pneumatic	Fill	Below		STC project files
51	Sprague-Roser Crossing of the Saline Creek	2013	Install conductor casing for HDD	Fort McMurray, Alberta	30 -760	197 -60						Inclined for HDD entry or exit	Michels case history
52	Highway Crossing under Paulista Metro Train Lines	1997	Built canopy arch with 25 rams under railroad	Mogi das Cruces, Brazil	32 -815	73 -22	40-foot	0.5 (13)	16-inch pneumatic	Soils	Above		Hammer-head case history
53	Montana Rail Link Culvert Project	2002	Install 10 culverts to accommodate Highway 200	Weeksville, Montana	36 -915	50 -15	20-foot	0.5 (13)		Mixed soils with gravel, cobbles, and boul-ders		2" to 3" below design grade	Werner, 2003
54	Transcanada Crossing of the Athabasca River	2007	Install conductor casing for HDD	Fort McCay, Alberta	36 -915	70 -21				Gravel and clay till		Inclined for HDD entry or exit	Michels case history
55	Kinder Morgan Crossing of the Michinsinlika River	2006	Install conductor casing for HDD	MacKenzie, British Columbia	36 -915	70 -21				Gravel and clay till		Inclined for HDD entry or exit	Michels case history

No.	Project Name	Year	Purpose	Location	Dia., in. (mm)		Drive Length, ft (m)		Casing Length/ Drive	Wall Thickness in (mm)	Hammer Type	Ground Type	Pipe Ram Position to Groundwater	Design Grade	Reference
56	Spectra Energy Transmission Pipeline	2015	Install conductor casing for HDD	Salem, Massachusetts	36	-915	76	-23		0.75 (19)				Inclined for HDD exit	Michels case history
57	Transcanada Crossing of the Ells River	2007	Install conductor casing for HDD	Fort McCay, Alberta	36	-915	80	-24						Inclined for HDD entry or exit	Michels case history
58	Transcanada Crossing of the Tar River	2007	Install conductor casing for HDD	Fort McCay, Alberta	36	-915	80	-24						Inclined for HDD entry or exit	Michels case history
59	Skykomish 2008 Remediation Railroad Utility Undercrossings	2008	Cross under railroad for twin utlilities	Skykomish, Washington	36	-915	90	-27	20-foot		14-inch pneumatic	Sand with gravel, cobbles, and boulders	Above		2009 Paper F-2-05
60	Colonial Pipeline Crossing of the Nansemond River	2017	Install conductor casing for HDD	Suffolk, Virginia	36	-915	92	-28						Inclined for HDD exit	Michels case history
61	Enbridge Crossing of Elleshmere Road	2008	Install conductor casing for HDD	Toronto, Ontario	36	-915	92	-28						Inclined for HDD entry or exit	Michels case history
62	Terasen-Kinder Morgan Crossing of the Clearwater River	2007	Install conductor casing for HDD	Fort McCay, Alberta	36	-915	92	-28						Inclined for HDD entry	Michels case history
63	Alliance Pipeline Project	2000	Install of 40 slick bores under roads, rails, creeks	Joilet, Illinois	36	-915	100	-31			18-inch pneumatic	Variable ground conditions	Above	Slick bores	TT case history
64	Voight's Creek Hatchery Fish Ladder	2014	Cross under Highway 162	Orting, Washington	36	-915	120	-37	20-foot	0.75 (19)		Silty sand and sandy silt	Above		2015 Paper TA-T4-01
65	TRAX Rail Line Crossings	2006	Install 3 casings for water mains under rail line	Salt Lake City, Utah	36	-915	120	-37	20-foot		18-inch pneumatic		Above	Pilot tube guided	TT case history

No.	Project Name	Year	Purpose	Location	Dia., in. (mm)	Drive Length, ft (m)	Casing Length/ Drive	Wall Thickness in (mm)	Hammer Type	Ground Type	Pipe Ram Position to Ground-water	Design Grade	Reference
66	Spectra Energy Transmission Pipeline	2015	Install conductor casing for HDD	Salem, Massachusetts	36 -915	120 -37		0.75 (19)				Inclined for HDD entry	Michels case history
67	Texas Eastern Transmission Pipeline Crossing of the Rahway River	2018	Install conductor casing for HDD	Linden, New Jersey	36 -915	127 -39		1.0 (25)			Below	Inclined for HDD entry or exit	Michels case history
68	Spectra Energy Transmission Pipeline	2015	Install conductor casing for HDD	Salem, Massachusetts	36 -915	134 -41		0.75 (19)				Inclined for HDD entry	Michels case history
69	Energy Transfer Company Pipeline	2018	Install conductor casing for HDD	Morgantown, Pennsylvania	36 -915	138 -42					Below	Inclined for HDD entry or exit	Michels case history
70	Greater Pine Island Water Line Crossing of the Matlacha River	2007	Install conductor casing for HDD	Matlacha, Florida	36 -915	140 -43						Inclined for HDD on south side	Michels case history
71	Surmont Phase 1 SAGD Project	2010	Install culvert under Highway 881	Fort McMurray, Alberta	36 -915	150 -46			24-inch pneumatic	Hard clay	Above	Pilot tube guided	TT case history
72	Duvernay Oil Crossing of the Wildhay River	2006	Install conductor casing for HDD	Hinton, Alberta	36 -915	150 -46				Clay with gravels and cobbles		Inclined for HDD entry or exit	Michels case history
73	Talisman Energy Crossing of the Torrens River	2006	Install conductor casing for HDD	Grand Prairie, Alberta	36 -915	150 -46				Clay with gravels and cobbles		Inclined for HDD entry or exit	Michels case history
74	Duke Energy Pipeline Crossing of Ohio River	2015	Install conductor casing for HDD	Covington, Kentucky	36 -915	158 -48						Inclined for HDD entry	Michels case history
75	Enterprise Products Pipeline Crossing of the Mississippi River	2017	Install conductor casing for HDD	Modeste, Louisiana	36 -915	164 -50		0.75 (19)			Below	Inclined for HDD exit	Michels case history

No.	Project Name	Year	Purpose	Location	Dia., in. (mm)	Drive Length, ft (m)	Casing Length/ Drive	Wall Thickness in (mm)	Hammer Type	Ground Type	Pipe Ram Position to Groundwater	Design Grade	Reference
76	NIPSCO Crossing of the North Saint Mary River	2016	Install conductor casing for HDD	Fort Wayne, Indiana	36 -915	165 -50		0.75 (19)				Inclined for HDD entry or exit	Michels case history
77	Explorer Pipeline Crossing of the Harbor Canal	2011	Install conductor casing for HDD	Gary, Indiana	36 -915	166 -51		0.75 (19)				Inclined for HDD entry	Michels case history
78	Enterprise Products Pipeline Crossing of the Mississippi River	2017	Install conductor casing for HDD	Modeste, Louisiana	36 -915	178 -54		0.75 (19)			Below	Inclined for HDD entry	Michels case history
79	LS 46 Regional Conveyance Project - 8-9	2016	Cross under wetland and stream	Covington, Washington	36 -915	180 -55	20-foot		24-inch pneumatic	Sandy gravel	Below	0.95%	2018 Paper MA-T4-04
80	TRAX Rail Line Crossing	2006	Install casing under TRAX rail line	Salt Lake City, Utah	36 -915	200 -61	20-foot		18-inch pneumatic		Above	Pilot tube guided	TT case history
81	Transcanada Crossing of the MacKay River	2007	Install conductor casing for HDD	Fort McCay, Alberta	36 -915	205 -63						Inclined for HDD entry or exit	Michels case history
82	Duvernay Oil Crossing of the Wildhay River	2006	Install conductor casing for HDD	Hinton, Alberta	36 -915	205 -63				Clay with gravels and cobbles		Inclined for HDD entry or exit	Michels case history
83	Transcanada Crossing of the Muskeg River	2007	Install conductor casing for HDD	Fort McCay, Alberta	36 -915	205 -63						Inclined for HDD entry or exit	Michels case history
84	Chervon Crossing of the Mississippi River	2009	Install conductor casing for HDD	Empire, Louisiana	36 -915	240 -73		0.75 (19)				Inclined for HDD entry or exit	Michels case history
85	Spectra Energy Transmission Pipeline	2015	Install conductor casing for HDD	Salem, Massachusetts	36 -915	247 -75		0.75 (19)				Inclined for HDD exit	Michels case history
86	Los Alamos Flood Drainage	2000	Cross under road for drainage	Los Alamos, New Mexico	36 -915	250 -76	40-foot		18-inch pneumatic	Dense cobble conglomerate			MOP First Edition

No.	Project Name	Year	Purpose	Location	Dia., in. (mm)		Drive Length, ft (m)		Casing Length/ Drive	Wall Thickness in (mm)	Hammer Type	Ground Type	Pipe Ram Position to Ground-water	Design Grade	Reference
87	Greater Pine Island Water Line Crossing of the Matlacha River	2007	Install conductor casing for HDD	Matlacha, Florida	36	-915	270	-82						Inclined for HDD on north side	Michels case history
88	Nicor Gas 36-inch Infrastructre Upgrade	2014	Extraction of existing gas line under I-88	Aurora, Illinois	36	-915	320	-98			24-inch pneumatic	Extrac-tion	Above		2014 Paper TM1-T4-03
89	Horizon Pipeline Project	2002	Install of 60 slick bores under roads, rails, creeks	Northern Illinois	36	-915	380	-116			18-inch pneumatic	Variable ground condi-tions	Above	Slick bores	TT case history
90	O'Hara Moderniza-tion Program		Cross under taxiway T with ductbank	Chicago, Illinois	36	-915	469	-143	20-foot			Satu-rated soft silt	Above	Pilot tube guided	2019 Paper MA-T4-01
91	O'Hara Moderniza-tion Program		Cross under taxiway SS with ductbank	Chicago, Illinois	36	-915	537	-164	20-foot			Satu-rated soft silt	Above	Pilot tube guided	2019 Paper MA-T4-01
92	Guelph Hydro Project		Cross under 4 lane highway	Guelph, Ontario	36	-915			20-foot			Sand and gravel with cobbles and boul-ders	Below		Waterloo Case History
93	Cross Country Gas Pipeline Installation	2004	Assist with 2,800-foot (854 m) long HDD pullback under the Navasota River	Normangee, Texas	36	-915					24-inch pneumatic		Below	HDD assist	Hammer-head case history
94	Gulfstream Pipeline Project	2002	Assist with double 4,700-foot (1,433 m) long HDD pullback	Bayou LaBatre, Alabama	36	-915					24-inch pneumatic	Marine soils	Below	HDD assist	TT case history

No.	Project Name	Year	Purpose	Location	Dia., in. (mm)	Drive Length, ft (m)	Casing Length/ Drive	Wall Thickness in (mm)	Hammer Type	Ground Type	Pipe Ram Position to Groundwater	Design Grade	Reference
95	Everett-Tulalip HDD Pipe Assist	2011	Assist with 2,690-foot (820 m) long HDD pullback	Everett, Washington	36 -915			0.75 (19)	24-inch pneumatic	Estuarine deposits and fine grained alluvium	Below	HDD assist	STC project files
96	SPU Pump Station 43 Emergency Pipeline	2019	Install conductor casing under railroad for HDD	Seattle, Washington	42 -1,070	40 -12	20-foot	0.5 (13)	20-inch pneumatic	Fill	Above	18 degree entry for HDD	STC project files
97	Spire Steel Pipeline Crossing of the Mississippi River	2019	Install conductor casing for HDD	Elsah, Illinois	42 -1,065	52 -16		1.0 (25)			Below	Inclined for HDD entry or exit	Michels case history
98	Spectra Energy Transmission Pipeline	2015	Install conductor casing for HDD	Salem, Massachusetts	42 -1,065	118 -36						Inclined for HDD entry	Michels case history
99	Surmont Phase 1 SAGD Project	2010	Install casing sleeve under Highway 881	Fort McMurray, Alberta	42 -1,070	120 -37			24-inch pneumatic	Hard clay	Above	Pilot tube guided	TT case history
100	HDD Assist by New Tech Construction	2019	Install conductor casing for HDD	Denver, Colorado	42 -1,070	120 -39			26-inch pneumatic		Above		2020 Paper
101	PG&E Crossing of the Latham Slough	2007	Install twin conductor casings for HDD	Sacramento, California	42 -1,070	120 -37						Inclined for HDD entry and exit	Michels case history
102	PG&E Crossing of the Old River	2007	Install conductor casing for HDD	Sacramento, California	42 -1,070	140 -43						Inclined for HDD entry or exit	Michels case history
103	HDD Assist by New Tech Construction	2019	Install conductor casing for HDD	Denver, Colorado	42 -1,070	148 -45			26-inch pneumatic		Above		2020 Paper
104	PG&E Crossing of the Middle River	2007	Install conductor casing for HDD	Sacramento, California	42 -1,070	160 -49						Inclined for HDD entry or exit	Michels case history
105	King County Crossing of I-405	2007	Install conductor casing for HDD	Bellevue, Washington	42 -1,070	168 -51						Inclined for HDD entry or exit	Michels case history

No.	Project Name	Year	Purpose	Location	Dia., in. (mm)	Drive Length, ft (m)	Casing Length/ Drive	Wall Thickness in (mm)	Hammer Type	Ground Type	Pipe Ram Position to Ground-water	Design Grade	Reference
106	Enbridge Norlite Crossing of the Athabasca River	2016	Install conductor casing for HDD	Wood Buffalo, Alberta	42 -1,065	190 -58		1.0 (25)				Inclined for HDD entry or exit	Michels case history
107	Spire Steel Pipeline Crossing of the Mississippi River	2019	Install conductor casing for HDD	Slim Island, Missouri	42 -1,065	230 -70		1.0 (25)			Below	Inclined for HDD entry or exit	Michels case history
108	Transcanada Crossing of the Wabasca River	2008	Install conductor casing for HDD	Wood Buffalo, Alberta	42 -1,070	360 -110		0.88 (22)				Inclined for HDD entry or exit	Michels case history
109	National Fuel Crossing of the Chemung River	2011	Install conductor casing for HDD	Corning, New York	42 -1,070	426 -130		0.50 (13)				Inclined for HDD entry or exit	Michels case history
110	CN Railway Crossing		Cross under CN railway	Toronto, Ontario	42 -1,070					Cobbles and boul-ders			Waterloo Case History
111	Enbridge Crossing of Highway 407	2011	Install conductor casing for HDD	Whitby, Ontario	48 -1,220	60 -18		0.50 (13)				Inclined for HDD entry	Michels case history
112	King County Crossing of I-405	2007	Install conductor casing for HDD	Bellevue, Washington	48 -1,220	80 -24						Inclined for HDD entry or exit	Michels case history
113	Duke Energy Crossing of the Merrimack River - Phase III	2003	Install conductor casing for HDD	Haverhill, Massachu-setts	48 -1,220	80 -24		1.00 (25)				Inclined for HDD entry or exit	Michels case history
114	Union Pacific Crossing	2006	Install casing under Union Pacific rail line	Salt Lake City, Utah	48 -1,220	90 -27	20-foot		18-inch pneumatic		Below		TT case history
115	Duke Energy Crossing of the Waters River	2003	Install conductor casing for HDD	Danvers, Massachu-setts	48 -1,220	90 -27		1.00 (25)				Inclined for HDD entry or exit	Michels case history
116	LS 46 Regional Conveyance Project - 7-6	2015	Depth linear installation	Covington, Washington	48 -1,220	100 -31	20-foot		24-inch pneumatic	Sandy gravel	Below	0.17%	2018 Paper MA-T4-04
117	King County Crossing of Southeast 6th Street	2007	Install conductor casing for HDD	Bellevue, Washington	48 -1,220	108 -33						Inclined for HDD entry or exit	Michels case history

No.	Project Name	Year	Purpose	Location	Dia., in. (mm)	Drive Length, ft (m)	Casing Length/ Drive	Wall Thickness in (mm)	Hammer Type	Ground Type	Pipe Ram Position to Ground-water	Design Grade	Reference
118	Transcanada-Keystone Crossing of the South Saskatchewan River	2012	Install conductor casing for HDD	Medicine Hat, Alberta	48 -1,220	110 -34						Inclined for HDD entry or exit	Michels case history
119	Texas Eastern Crossing of East First Street	2013	Install conductor casing for HDD	Bayonne, New Jersey	48 -1,220	113 -34		0.75 (19)				Inclined for HDD exit	Michels case history
120	Chelsea River Electrical Ductbanks	2011	Install 3 conduit casings for HDD crossings	Boston, Massachusetts	48 -1,220	120 -37	40-foot	0.75 (19)	24-inch pneumatic	Contaminated sand	Below	16 degree entry for HDD	2012 Paper A-1-03
121	Voight's Creek Hatchery Fish Ladder	2014	Cross under Highway 162	Orting, Washington	48 -1,220	120 -37	20-foot	0.75 (19)		Silty sand and sandy silt	Above		2015 Paper TA-T4-01
122	Duke Energy Crossing of the Weymouth River	2003	Install conductor casing for HDD	Weymouth, Massachusetts	48 -1,220	120 -37		1.00 (25)				Inclined for HDD exit	Michels case history
123	Williams Gas Pipeline Crossing of the White River	2004	Install conductor casing for HDD	Auburn, Washington	48 -1,220	122 -37						Inclined for HDD entry or exit	Michels case history
124	Transcanada Pipeline Crossing of Beaver Lake	2015	Install conductor casing for HDD		48 -1,220	129 -39						Inclined for HDD entry or exit	Michels case history
125	Montgomery County Sewer Replacement Project	2008	Install casing under the Stillwater River	Dayton, Ohio	48 -1,220	130 -40	43-foot	0.63 (16)	23-inch pneumatic	Clay, sand, gravel, cobbles, and boulders	Below	0.19% design grade	Hammerhead case history
126	Sprague-Roser Crossing of the Saline Creek	2013	Install conductor casing for HDD	Fort McMurray, Alberta	48 -1,220	130 -40						Inclined for HDD entry or exit	Michels case history

No.	Project Name	Year	Purpose	Location	Dia., in. (mm)	Drive Length, ft (m)	Casing Length/ Drive	Wall Thickness in (mm)	Hammer Type	Ground Type	Pipe Ram Position to Ground-water	Design Grade	Reference
127	Duke Energy Crossing of Beverly Harbor	2003	Install conductor casing for HDD	Salem, Massachu-setts	48 -1,220	140 -43		1.00 (25)				Inclined for HDD entry or exit	Michels case history
128	Ford Island to Landing C	2017	Install conductor casing for HDD	Honolulu, Hawaii	48 -1,220	157 -48		0.75 (19)			Below	Inclined for HDD entry or exit	Michels case history
129	Enrergy Transfer Company Crossing of the Missouri River	2017	Install conductor casing for HDD	Cannonball, North Dakota	48 -1,220	159 -48		0.75 (19)				Inclined for HDD entry or exit	Michels case history
130	King County Crossing of Southeast 6th Street	2007	Install conductor casing for HDD	Bellevue, Washington	48 -1,220	160 -49						Inclined for HDD entry or exit	Michels case history
131	Georgia Power Crossing of the Chattahoochie River	2009	Install conductor casing for HDD	Atlanta, Georgia	48 -1,220	160 -49		0.75 (19)				Inclined for HDD entry	Michels case history
132	Texas Eastern Crossing of Bayonne Inlet	2013	Install conductor casing for HDD	Bayonne, New Jersey	48 -1,220	161 -49						Inclined for HDD exit	Michels case history
133	Kinder Morgan Gas Pipeline Crossing of the Bennekill Stream	2013	Install conductor casing for HDD	Milford, New Jersey	48 -1,220	163 -50		0.75 (19)				Inclined for HDD entry or exit	Michels case history
134	Texas Eastern Crossing of Bayonne Inlet	2013	Install conductor casing for HDD	Bayonne, New Jersey	48 -1,220	181 -55						Inclined for HDD entry	Michels case history
135	FEMA Culvert Improvement Project	2013	Install 10 new culverts under railroad	Jefferson County, Colorado	48 -1,220	200 -61	20-foot			Fill	Above		2015 Paper TA-T4-02
136	AK DOT	2009	Install drainage culvert /salmon habitat enhancement	Fairbanks, Alaska	48 -1,220	200 -61		1.0 (25)	150kJ / hydraulic	Glacial till	Below		Crossing Group case history
137	City of Coronado Crossing of San Diego Bay	2007	Install conductor casing for HDD	San Diego, California	48 -1,220	200 -61						Inclined for HDD entry or exit	Michels case history
138	Otay Water District's Crossing of the Otay River	2006	Install conductor casing for HDD	Otay, California	48 -1,220	210 -64						Inclined for HDD entry or exit	Michels case history

No.	Project Name	Year	Purpose	Location	Dia., in. (mm)	Drive Length, ft (m)	Casing Length/ Drive	Wall Thickness in (mm)	Hammer Type	Ground Type	Pipe Ram Position to Ground-water	Design Grade	Reference
139	Williams Atlantic Pipeline Crossing of the Susquehanna River	2018	Install conductor casing for HDD	Hoban Heights, Pennsylva-nia	48 -1,220	232 -71		1.0 (25)			Below	Inclined for HDD entry or exit	Michels case history
140	FEMA Culvert Improvement Project	2013	Install 4 new culverts under railroad	Jefferson County, Colorado	48 -1,220	235 -72	20-foot			Fill	Above		2015 Paper TA-T4-02
141	Texas Eastern Crossing of Kill Van Kull	2013	Install conductor casing for HDD	Bergen Point New Jersey	48 -1,220	236 -72		1.00 (25)				Inclined for HDD entry	Michels case history
142	Williams Gas Pipeline Crossing of McClane Park	2002	Install conductor casing for HDD	Tumwater, Washington	48 -1,220	240 -73		1.00 (25)				Inclined for HDD entry or exit	Michels case history
143	Texas Eastern Crossing of East First Street	2013	Install conductor casing for HDD	Bayonne, New Jersey	48 -1,220	241 -73		0.75 (19)				Inclined for HDD entry	Michels case history
144	Fairwood Emergency Culvert Repair	2009	Install replacement culvert	Fairwood, Washington	48 -1,220	249 -76	20-foot	0.75 (19)		Fill into till	Above	3.80%	STC project files
145	O'Hara Moderniza-tion Program		Cross under taxiway T with watermain	Chicago, Illinois	48 -1,220	260 -79	20-foot			Satu-rated soft silt	Above	Pilot tube guided	2019 Paper MA-T4-01
146	Georgia Power Crossing of I-285 North	2009	Install conductor casing for HDD	Atlanta, Georgia	48 -1,220	260 -79		0.75 (19)				Inclined for HDD entry or exit	Michels case history
147	Texas Eastern Crossing of Merseles Street	2013	Install conductor casing for HDD	Hamilton Park, New Jersey	48 -1,220	294 -90						Inclined for HDD exit	Michels case history
148	LS 46 Regional Conveyance Project - 5-6	2016	Cross under State Highway 18	Covington, Washington	48 -1,220	299 -91	20-foot		24-inch pneumatic	Gravelly sand	Below	3.37%	2018 Paper MA-T4-04
149	Encana Oil and Gas Crossing of the Colorado River	2011	Install conductor casing for HDD	Parachute, Colorado	48 -1,220	300 -91		0.75 (19)				Inclined for HDD entry or exit	Michels case history
150	Guardian Pipeline Crossing of the Rock River	2008	Install conductor casing for HDD	Dodge County, Wisconsin	48 -1,220	300 -91						Inclined for HDD entry or exit	Michels case history

No.	Project Name	Year	Purpose	Location	Dia., in. (mm)	Drive Length, ft (m)	Casing Length/ Drive	Wall Thickness in (mm)	Hammer Type	Ground Type	Pipe Ram Position to Ground-water	Design Grade	Reference
151	National Fuel Crossing of the Chemung River	2011	Install conductor casing for HDD	Corning, New York	48 -1,220	322 -98		1.00 (25)				Inclined for HDD entry or exit	Michels case history
152	Texas Eastern Crossing of the Hudson River	2012	Install conductor casing for HDD	Newport Village, New Jersey	48 -1,220	377 -115						Inclined for HDD entry	Michels case history
153	O'Hara Moderniza-tion Program		Cross under taxiway R with watermain	Chicago, Illinois	48 -1,220	380 -116	20-foot			Satu-rated soft silt	Above	Pilot tube guided	2019 Paper MA-T4-01
154	Inergy Crossing of the Susquehanna River	2012	Install conductor casing for HDD	Wyalusing, Pennsylva-nia	48 -1,220	384 -117		0.75 (19)				Inclined for HDD entry or exit	Michels case history
155	Texas Eastern Crossing of Merseles Street	2013	Install conductor casing for HDD	Hamilton Park, New Jersey	48 -1,220	404 -123						Inclined for HDD entry	Michels case history
156	Duke Energy Crossing of Georges Island	2003	Install conductor casing for HDD	Salem, Massachu-setts	48 -1,220	810 -247		1.00 (25)				Inclined for HDD entry or exit	Michels case history
157	Surmont Phase 1 SAGD Project	2010	Install 2 casing sleeves under Highway 881	Fort McMurray, Alberta	52 -1,320	120 -37			24-inch pneumatic	Hard clay	Above	Pilot tube guided	TT case history
158	Ma Wan Channel HDD Water Line	1999	Install conductor casing under Castle Peak Road	Angler's Beach, Hong Kong	54 -1,370	66 -20	20-foot		23-inch pneumatic	Soils	Below	20 degree entry for HDD	Hammer-head case history
159	Transcanada-Key-stone Crossing of the Red Deer River	2012	Install conductor casing for HDD	Cavendish, Alberta	54 -1,370	66 -20		0.75 (19)				Inclined for HDD entry or exit	Michels case history
160	Inergy Crossing of the Susquehanna River	2012	Install conductor casing for HDD	Wyalusing, Pennsylva-nia	54 -1,370	97 -30		0.75 (19)				Inclined for HDD entry or exit	Michels case history
161	Enbridge Pipeline Crossing of Highway 12 and Amtrak	2018	Install conductor casing for HDD	Rawson-ville, Michgan	54 -1,370	107 -33		1.0 (25)			Below	Inclined for HDD entry or exit	Michels case history
162	Sprague-Roser Crossing of the Saline Creek	2013	Install conductor casing for HDD	Fort McMurray, Alberta	54 -1,370	131 -40		1.00 (25)				Inclined for HDD entry or exit	Michels case history

No.	Project Name	Year	Purpose	Location	Dia., in. (mm)		Drive Length, ft (m)		Casing Length/ Drive	Wall Thickness in (mm)	Hammer Type	Ground Type	Pipe Ram Position to Ground-water	Design Grade	Reference
163	PSE&G Crossing of the Hackensack River	2012	Install conductor casing for HDD	Secaucus, New Jersey	54	-1,370	140	-43						Inclined for HDD entry or exit	Michels case history
164	Everett-Tulalip Segment 2 Pipeline	2016	Cross under railroad	Tulalip, Washington	54	-1,370	170	-52	20-foot	0.75 (19)	24-inch pneumatic	Estua-rine deposits	Below	0.95%	STC project files
165	Spokane Water Line Replacement Project	1999	Install casing sleeve under Burlington Northern Railroad	Spokane, Washington	54	-1,370	180	-55	20-foot	0.72 (18)	23-inch pneumatic	Glacial tills	Above	0.25% design grade	Hammer-head case history
166	City of Charleston Crossing of the Wando River	2007	Install conductor casing for HDD	Charleston, South Carolina	54	-1,370	200	-61						Inclined for HDD entry or exit	Michels case history
167	Trans Canada Pipeline Crossing of the Ohio River	2017	Install conductor casing for HDD	Westmore-land, West Virginia	54	-1,370	272	-83					Below	Inclined for HDD entry or exit	Michels case history
168	Willamette River Crossing	2014	Install conductor casing for HDD	Gladstone, Oregon	54	-1,370	393	-120	20-foot	1.0 (25)	24-inch pneumatic	Gravels and cobbles	Below	9 degree entry for HDD	STC project files
169	Lake Oswego-Tigard Willamette River Crossing	2014	Install conductor casing for HDD	Lake Oswego, Oregon	54	-1,370	393	-120		1.00 (25)				Inclined for HDD entry or exit	Michels case history
170	Lake Scanton Finished Water Tunnel	2018	Install conductor casing for HDD	Scranton, Pennsylva-nia	60	-1,525	40	-12		0.75 (19)			Below	Inclined for HDD entry or exit	Michels case history
171	Williams Transconti-nental Gas Pipeline Crossing of I-80	2018	Install conductor casing for HDD	Blooms-burg, Pennsylva-nia	60	-1,525	40	-12		1.0 (25)			Below	Inclined for HDD entry or exit	Michels case history
172	Transcanada Pipeline Crossing of the McKay River	2016	Install conductor casing for HDD	Fort McMurray, Alberta	60	-1,525	63	-19		1.0 (25)				Inclined for HDD entry or exit	Michels case history
173	Transcanada Pipeline Crossing of the Dover River	2016	Install conductor casing for HDD	Fort McMurray, Alberta	60	-1,525	66	-20		1.0 (25)				Inclined for HDD entry or exit	Michels case history

No.	Project Name	Year	Purpose	Location	Dia., in. (mm)	Drive Length, ft (m)	Casing Length/ Drive	Wall Thickness in (mm)	Hammer Type	Ground Type	Pipe Ram Position to Ground-water	Design Grade	Reference
174	Transcanada Pipeline Crossing of Ruth Lake	2015	Install conductor casing for HDD		60 -1,525	69 -21		1.00 (25)				Inclined for HDD entry or exit	Michels case history
175	Union Pacific Crossing	2006	Install casing under Union Pacific rail line	Salt Lake City, Utah	60 -1,525	80 -24	20-foot		18-inch pneumatic		Above		TT case history
176	Transcanada-Keystone Crossing of the South Saskatchewan River	2012	Install conductor casing for HDD	Medicine Hat, Alberta	60 -1,525	86 -26						Inclined for HDD entry or exit	Michels case history
177	CN Rail Line Culvert 71.30 Replacement	2013	Install twin culverts under railroad	Askum, Illinois	60 -1,525	90 -27	20-foot		24-inch pneumatic	Soft ground	Above	1.1% design grade	Hammer-head case history
178	Nexus Gas Pipeline Crossing of the Tuscarawas River	2018	Install conductor casing for HDD	Greensburg, Ohio	60 -1,525	90 -27		0.75 (19)			Below	Inclined for HDD entry or exit	Michels case history
179	Enbridge Crossing of the Athabasca River	2011	Install conductor casing for HDD	Fort McMurray, Alberta	60 -1,525	90 -27		1.00 (25)				Inclined for HDD entry or exit	Michels case history
180	Everett-Tulalip Segment 2 Pipeline	2016	Install conductor casing for HDD	Tulalip, Washington	60 -1,525	96 -29	20-foot	0.75 (19)	24-inch pneumatic	Estua-rine deposits	Below	12 degree exit for HDD	STC project files
181	Transcanada Pipeline Crossing of Major MacKenzie Road	2016	Install conductor casing for HDD	Ontario	60 -1,525	98 -30		0.81 (21)				Inclined for HDD entry or exit	Michels case history
182	Burlington Northern Railroad Storm Water Improvements	2008	Install culvert under railroad	Rockville, Minnesota	60 -1,525	100 -31	20-foot		24-inch pneumatic	Loose sand and cobbles	Above		Hammer-head case history
183	Jacksonville Electric Crossing of the Saint John River	2012	Install conductor casing for HDD	Jackson-ville, Florida	60 -1,525	100 -30		1.00 (25)				Inclined for HDD entry	Michels case history
184	Guam Water Beach Approach at Hagatna	2008	Install conductor casing for HDD	Guam	60 -1,525	100 -30		0.75 (19)				Inclined for HDD entry or exit	Michels case history

No.	Project Name	Year	Purpose	Location	Dia., in. (mm)	Drive Length, ft (m)	Casing Length/ Drive	Wall Thickness in (mm)	Hammer Type	Ground Type	Pipe Ram Position to Ground-water	Design Grade	Reference
185	Transcanada Pipeline Crossing of Beaver Lake	2015	Install conductor casing for HDD	Fort McMurray, Alberta	60 -1,525	107 -33		1.00 (25)				Inclined for HDD entry or exit	Michels case history
186	Williams Transcontinental Gas Pipeline Crossing of the Susquehanna River	2018	Install conductor casing for HDD	Catawissa, Pennsylvania	60 -1,525	114 -35		1.0 (25)			Below	Inclined for HDD entry or exit	Michels case history
187	Texas Eastern Crossing of Kill Van Kull	2013	Install conductor casing for HDD	Bergen Point, New York	60 -1,525	120 -37						Inclined for HDD exit	Michels case history
188	Harbour Resource Partners' Pipeline Crossing of Victoria Harbour	2018	Install conductor casing for HDD	Victoria, Bristish Columbia	60 -1,525	122 -37		1.0 (25)			Below	Inclined for HDD entry or exit	Michels case history
189	Energy Transfer Company Crossing of the Mississippi River	2016	Install conductor casing for HDD	Sandusky, Iowa	60 -1,525	136 -41		1.00 (25)				Inclined for HDD entry	Michels case history
190	Everett-Tulalip Segment 2 Pipeline	2016	Install conductor casing for HDD	Tulalip, Washington	60 -1,525	137 -42	20-foot	0.75 (19)	24-inch pneumatic	Estuarine deposits	Below	14 degree entry for HDD	STC project files
191	Trans Canada Pipeline Crossing of the Athabasca River	2015	Install conductor casing for HDD	Fort McMurray, Alberta	60 -1,525	140 -43		1.0 (25)				Inclined for HDD entry or exit	Michels case history
192	North Creek Interceptor	2015	Cross under a creek for sewer	Bothell, Washington	60 -1,525	145 -44	20-foot	1.0 (25)	32-inch pneumatic	Sand and gravel alluvium	Below	0.0025	STC project files
193	Enbridge Norlite Crossing of the Clearwater River	2015	Install conductor casing for HDD	Wood Buffalo, Alberta	60 -1,525	145 -44		0.75 (19)				Inclined for HDD entry or exit	Michels case history
194	Colorado DOT Culvert Replacement Project		Install twin culverts under Highway 36	Estes Park, Colorado	60 -1,525	150 -46	20-foot		26-inch pneumatic	Ground with cobbles	Above		2019 Paper MA-T4-03

No.	Project Name	Year	Purpose	Location	Dia., in. (mm)	Drive Length, ft (m)	Casing Length/ Drive	Wall Thickness in (mm)	Hammer Type	Ground Type	Pipe Ram Position to Ground-water	Design Grade	Reference
195	Texas Eastern Crossing of I-95 and Tidal Wetlands	2013	Install conductor casing for HDD	Linden, New Jersey	60 -1,525	151 -46						Inclined for HDD entry or exit	Michels case history
196	Energy Transfer Company Crossing of the Mississippi River	2016	Install conductor casing for HDD	Nauvoo, Illinois	60 -1,525	153 -47		1.00 (25)				Inclined for HDD exit	Michels case history
197	Williams Transconti-nental Gas Pipeline Crossing of the Susquehanna River	2018	Install conductor casing for HDD	Catawissa, Pennsylva-nia	60 -1,525	160 -49		1.0 (25)			Below	Inclined for HDD entry or exit	Michels case history
198	Lake Oswego-Tigard Willamette River Crossing	2014	Install conductor casing for HDD	Lake Oswego, Oregon	60 -1.525	166 -51		1.00 (25)				Inclined for HDD entry or exit	Michels case history
199	Transcanada Pipeline Crossing of Major MacKenzie Road	2016	Install conductor casing for HDD	Ontario	60 -1,525	171 -52		0.81 (21)				Inclined for HDD entry or exit	Michels case history
200	Lake Oswego-Tigard Willamette River Crossing	2014	Install conductor casing for HDD	Lake Oswego, Oregon	60 -1,525	174 -53		1.00 (25)				Inclined for HDD entry or exit	Michels case history
201	Willamette River Crossing	2014	Install conductor casing for HDD	Gladstone, Oregon	60 -1,525	175 -53	20-foot	1.0 (25)	24-inch pneumatic	Gravels and cobbles	Below	9 degree entry for HDD	STC project files
202	CN Rail Line Culvert 359.7 Replacement	2013	Install twin culverts under railroad	Cairo, Illinois	60 -1,525	180 -55	30-foot	0.875 (22)	24-inch pneumatic	Soupy ground with stumps and piers	Above	2% design grade	Hammer-head case history
203	Guam Water Beach Approach at Northern Shore	2009	Install conductor casing for HDD	Guam	60 -1,525	180 -55		0.75 (19)				Inclined for HDD entry or exit	Michels case history
204	Jacksonville Electric Saint John River Crossing	2016	Install conductor casing for HDD	Jackson-ville, Florida	60 -1,525	186 -57		1.0 (25)				Inclined for HDD entry or exit	Michels case history

No.	Project Name	Year	Purpose	Location	Dia., in. (mm)	Drive Length, ft (m)	Casing Length/ Drive	Wall Thickness in (mm)	Hammer Type	Ground Type	Pipe Ram Position to Ground-water	Design Grade	Reference
205	Texas Eastern Crossing of I-95 and Tidal Wetlands	2013	Install conductor casing for HDD	Linden, New Jersey	60 -1,525	199 -61						Inclined for HDD entry or exit	Michels case history
206	AK DOT	2009	Install drainage culvert /salmon habitat enhancement	Fairbanks, Alaska	60 -1,525	200 -61		1.0 (25)	150kJ / Hydraulic	Glacial till	Below		Crossing Group case history
207	Enbridge Pipeline Crossing of the Christina River	2016	Install conductor casing for HDD	Fort McMurray, Alberta	60 -1,525	200 -61		1.0 (25)				Inclined for HDD entry or exit	Michels case history
208	Jacksonville Electric Crossing of the Saint John River	2012	Install conductor casing for HDD	Jackson-ville, Florida	60 -1,525	200 -61		1.00 (25)				Inclined for HDD exit	Michels case history
209	Pouce Coupe River Crossing	2019	Install conductor casing for HDD	Pouce Coupe, British Columbia	60 -1,525	210 -64		1.0 (25)			Below	Inclined for HDD entry or exit	Michels case history
210	Trans Canada Pipeline Crossing of the Ohio River	2017	Install conductor casing for HDD	Burlington, Ohio	60 -1,525	218 -66		0.75 (19)			Below	Inclined for HDD entry or exit	Michels case history
211	Texas Eastern Crossing of Tidal Wetlands	2013	Install conductor casing for HDD	Staten Island, New Jersey	60 -1,525	235 -72						Inclined for HDD exit	Michels case history
212	Everett-Tulalip Segment 3 Pipeline	2007	Install conductor casing for HDD	Tulalip, Washington	60 -1,525	240 -73	20-foot	0.75 (19)	24-inch pneumatic	Estua-rine deposits	Below		STC project files
213	National Fuel Crossing of the Chemung River	2011	Install conductor casing for HDD	Corning, New York	60 -1,525	240 -73		1.00 (25)				Inclined for HDD entry or exit	Michels case history
214	Texas Eastern Crossing of Arthur Kill	2013	Install conductor casing for HDD	Staten Island, New York	60 -1,525	244 -74		1.00 (25)				Inclined for HDD entry	Michels case history
215	Texas Eastern Crossing of Tidal Wetlands	2013	Install conductor casing for HDD	Staten Island, New Jersey	60 -1,525	248 -76		1.00 (25)				Inclined for HDD entry	Michels case history
216	Enbridge Crossing of the Kishwaukee River	2008	Install conductor casing for HDD	Belvidere, Illinois	60 -1,525	320 -128						Inclined for HDD entry or exit	Michels case history

No.	Project Name	Year	Purpose	Location	Dia., in. (mm)	Drive Length, ft (m)	Casing Length/ Drive	Wall Thickness in (mm)	Hammer Type	Ground Type	Pipe Ram Position to Ground-water	Design Grade	Reference
217	San Jacinto River Authority Raw Water Canal Project	2003	Install twin casings under Union Pacific Railroad	Highland, Texas	66 -1,675	40 -12	40-foot		24-inch pneumatic		Above		TT case history
218	Inter Pipeline Limited Crossing of the Vermillion River	2013	Install conductor casing for HDD	Vermillion, Alberta	66 -1,675	40 -12		0.63 (16)				Inclined for HDD entry or exit	Michels case history
219	Inter Pipeline Limited Crossing of the Saskatchewan River	2013	Install conductor casing for HDD	Elk Point, Alberta	66 -1,675	87 -27		1.00 (25)				Inclined for HDD entry or exit	Michels case history
220	The Crossing Group	2019	Install conductor casing for HDD	British Columbia	66 -1,675	90 -27			34-inch pneumatic	Lime-stone	Above	12 degree entry for HDD	2019 Paper No-Dig North
221	VTrans FEMA Flooding Repairs	2018	Install twin culverts under railway	Ferrisburgh, Vermont	72 -1,830	50 -15	20-foot	1.0 (25)	24-inch pneumatic	Clay soils	Above	Pilot tube guided	2019 Paper TA-T1-05
222	Hoffmans Bluff	2015	Install twin drainage culverts	Kamloops, British Columbia	72 -1,830	70 -21		1.0 (25)	90kJ / hydraulic	Silt	Below		Crossing Group case history
223	North Creek Interceptor	2015	Cross under a creek for sewer	Bothell, Washington	72 -1,830	84 -26	20-foot	1.0 (25)	32-inch pneumatic	Sand and gravel allu-vium	Below	0.25%	STC project files
224	Ontario Culvert Replacement Project		Swallow twin culvers under Highway 634	Smooth Rock Falls, Ontario	72 -1,830	100 -31	40-foot		24-inch pneumatic		Above		2016 Paper MA-T4-04
225	Spanish Fork Santaquin Pipeline	2017	Cross under highway for water transmission	Spanish Fork, Utah	72 -1,830	107 -33	40-foot	1.0 (25)	24-inch pneumatic	Cobbly gravel, sand, silt, and clay with boul-ders	Above		Central Utah Water Conser-vancy District

No.	Project Name	Year	Purpose	Location	Dia., in. (mm)	Drive Length, ft (m)	Casing Length/ Drive	Wall Thickness in (mm)	Hammer Type	Ground Type	Pipe Ram Position to Ground-water	Design Grade	Reference
226	Spanish Fork Santaquin Pipeline	2017	Cross under railroad for water transmission	Spanish Fork, Utah	72 -1,830	119 -36	40-foot	1.0 (25)	24-inch pneumatic	Cobbly gravel, sand, silt, and clay with boul-ders	Above		Central Utah Water Conser-vancy District
227	Vermont Bridge Project VT78 Bridget 10 SCRP (12)	2017	Install culvert under highway	Highgate, Vermont	72 -1,830	131 -40	20-foot	0.75 (19)	24-inch pneumatic	Silty clay and silt cobbles, and boul-ders	Within		Vermont DOT
228	North Creek Interceptor	2015	Cross under a road for sewer	Bothell, Washington	72 -1,830	150 -46	20-foot	1.0 (25)	32-inch pneumatic	Glacial sand and gravels with cobbles and boul-ders	Below	0.21%	STC project files
229	North Creek Interceptor	2015	Cross under a pond for sewer	Bothell, Washington	72 -1,830	170 -52	20-foot	1.0 (25)	32-inch pneumatic	Glacial sand and gravel with cobbles and boul-ders	Below	0.25%	STC project files
230	1-393 Culvert Replacement Project	2016	Install culvert under highway	Concord, New Hampshire	72 -1,830	228 -70	20-foot	1.0 (25)	24-inch pneumatic	Dens-esand and silt embank-ment fill	Within		New Hampshire DOT

No.	Project Name	Year	Purpose	Location	Dia., in. (mm)	Drive Length, ft (m)	Casing Length/ Drive	Wall Thickness in (mm)	Hammer Type	Ground Type	Pipe Ram Position to Ground-water	Design Grade	Reference
231	Caltrans Culvert Replacement Project	2016	Swallow culvert under I-80	Colfax, California	72 -1,830	250 -76	40-foot	1.25 (32)	34-inch pneumatic	Soft ground	Above	6.8 degrees upward	2018 Paper MA-T4-03
232	Highway 238 to Jackson Project	2000	Cross under Highway 99	Medford, Oregon	80 -2,030	80 -24	20-foot		24-inch pneumatic			0.16" of design grade	TT case history
233	Meyers Materials Gravel Quarry Expansion	2003	Install casing under Highway 120	McHenry, Illinois	82 -2,085	150 -46	50-foot		32-inch pneumatic		Above	1.5% downgrade	TT case history
234	Highway 160 Emergency Culvert Replacement Project	2010	Swallow 60-inch CMP culvert	Pagosa Springs, Colorado	84 -2,135	80 -24	20-foot		24-inch pneumatic	Soils	Above		2011 Paper C-4-04
235	Olympus Meadows Trunk Water & Sewer	2015	Cross under Silver Creek and 196th Street	Bothell, Washington	84 -2,135	86 -26	20-foot	1.25 (32)	24-inch pneumatic	Cobble boulder glacial outwash	Below	0.70%	2018 Paper MA-T4-02
236	Olympus Meadows Trunk Water & Sewer	2016	Cross under SR-227	Bothell, Washington	84 -2,135	151 -46	20-foot	1.25 (32)	24-inch pneumatic	Cobble boulder glacial outwash	Below	0.70%	2018 Paper MA-T4-02
237	Middle River Intake & Pump Station	2010	Retract stuck MTBM and casing	Bethyl Island, California	96 -2,440	32 -10	10-foot	0.75 (19)	24-inch pneumatic	Fine grained allu-vium	Below	N/A	MJA
238	CN/Unnamed Creek	2017	Install drainage culvert	Field, British Columbia	96 -2,440	90 -27		1.0 (25)	90kJ / hydraulic	Gravel	Above		Crossing Group case history
239	Highway 21 Culvert Replacement Project	2010	Install new culvert under highway	Goderich, Ontario	96 -2,440	128 -39	40-foot	1.0 (25)	32-inch pneumatic	Sandy silt and Silty sand w/ gravel	Above		2014 Paper TM1-T4-04
240	Tyee passing lane	2017	Install drainage culvert	Smithers, British Columbia	96 -2,440	150 -46		1.25 (32)	90kJ / hydraulic	Fill	Above		Crossing Group case history

No.	Project Name	Year	Purpose	Location	Dia., in. (mm)	Drive Length, ft (m)	Casing Length/ Drive	Wall Thickness in (mm)	Hammer Type	Ground Type	Pipe Ram Position to Groundwater	Design Grade	Reference
241	Lexington Rd	2016	Install drainage culvert	Williams Lake, British Columbia	96 -2,440	160 -49		1.0 (25)	90kJ / hydraulic	Gravel	Below		Crossing Group case history
242	Rio Tinto/Unnamed Creek	2018	Install drainage culvert	Setp-Illes, Quebec	96 -2,440	250 -76		1.0 (25)	90kJ / hydraulic	Fill	Below		Crossing Group case history
243	Sioux Fall Outfall Sewer Project	2017	Cross under Interstate 229	Sioux Falls, South Dakota	96 -2,440	350 -107				Glacial tills and allu-vium	Above	0.15%	2019 Paper TM1-T4-03
244	Jarvis River Crossing Porject	2018	Cross under Highway 61	Neebing, Ontario	96 -2,440			1.0 (25)	24-inch pneumatic		Above		2019 Paper TM1-T4-02
245	Magnolia Stormwa-ter Project	2015	Cross under railroad	El Paso, Texas	98 -2,490	105 -32	20-foot		32-inch pneumatic	Uniform clean sand	Above	Pilot tube guided	2019 Paper MM-T1-02
246	Culvert Upsize at Mile Post 29.2	2015	Swallow culvert under Highway 1	Timber Gulch, California	108 -2,745	88 -27	20-foot	1.25 (32)	34-inch pneumatic	Clayey and silty gravel	Above	9.4% downgrade	2016 Paper MA-T4-01
247	Port Mann/Latimer Creek 1	2011	Install drainage culvert /salmon habitat enhancement	Langley, British Columbia	120 -3,050	130 -40		1.0 (25)	150kJ / hydraulic	Glacial till	Below		Crossing Group case history
248	Port Mann/Latimer Creek 2	2012	Install drainage culvert /salmon habitat enhancement	Langley, British Columbia	120 -3,050	165 -50		1.0 (25)	150kJ / hydraulic	Glacial till	Below		Crossing Group case history
249	Ryan Creek Fish Passage Mitigation Project	2017	Install twin culverts under Highway 101	Willits, California	120 -3,050	170 -52		1.5 (38)	150 kJ hydraulic	Silt and sand with gravel	At the same level	2.9% slope (5'/170')	2018 Paper MA-T4-01
250	AK DOT	2009	Install drainage culvert /salmon habitat enhancement	Fairbanks, Alaska	120 -3,050	180 -55		1.0 (25)	150kJ / hydraulic	Glacial till	Below		Crossing Group case history
251	Bjorgen Creek	2010	Install drainage culvert /salmon habitat enhancement	Washington	120 -3,050	260 -79			200kJ / hydraulic	Glacial till	Below		Crossing Group case history

No.	Project Name	Year	Purpose	Location	Dia., in. (mm)		Drive Length, ft (m)		Casing Length/ Drive	Wall Thickness in (mm)	Hammer Type	Ground Type	Pipe Ram Position to Ground-water	Design Grade	Reference
252	Port Mann/Leoran Brook	2013	Install drainage culvert /salmon habitat enhancement	Surrey, British Columbia	120	-3,050	280	-85		1.0 (25)	150kJ / hydraulic	Glacial till	Below		Crossing Group case history
253	Peoria & Pekin Union Railroad Bike Path Project	2000	Cross under railroad for bike path	East Peoria, Illinois	122	-3100	50	-15	10-foot		24-inch pneumatic			0.5" of design grade	TT case history
254	Townsend Creek	2019	Install drainage culvert	Fort St. John, British Columbia	126	-3,200	200	-61		1.0 (25)	90kJ / hydraulic	Fill	Below		Crossing Group case history
255	CN/Unnamed Creek	2016	Install drainage culvert	Banff, Alberta	132	-3,355	82	-25		1.0 (25)	90kJ / hydraulic	Gravel	Above		Crossing Group case history
256	OHSU Block 28 and 29 Design Build Tunnel	2016	Pedestrian crossing between two buildings	Portland, Oregon	144	-3660	70	-21	8-foot	1.25 (32)	90kJ / hydraulic	Struc-tural fill	Below		2020 Paper
257	AK DOT	2009	Install drainage culvert /salmon habitat enhancement	Fairbanks, Alaska	144	-3,660	180	-55		1.0 (25)	150kJ / hydraulic	Glacial till	Below		Crossing Group case history
258	Highway 63 Widening Project	2013	Install culvert under Highway 63	Fort McMurray, Alberta	144	-3,660	237	-72			32-inch pneumatic	Heavy moist clay	Above		2014 Paper TM1-T4-05
259	Bjorgen Creek	2010	Install drainage culvert /salmon habitat enhancement	Washington	144	-3,660	270	-82			200kJ / hydraulic	Glacial till	Below		Crossing Group case history
260	Pernham Creek	2007	Install drainage culvert /salmon habitat enhancement	Oregon	144	-3,660	275	-84		1.25 (32)	280kJ / hydraulic	Glacial till	Below		Crossing Group case history
261	Iowa Interstate Railroad Crossing	2004	Cross under railroad for bike path	Altoona, Iowa	147	-3,734	60	-18	20-foot	1.5 (38)	24-inch pneumatic		Above	2.2% downgrade	MOP First Edition

GLOSSARY

Acceleration. Change in velocity per change in time.

Adaptor. Any device that connects the hammer to the casing, which can include cones, collets, and collars.

Advance Rate. Forward progress over a given period of time, which includes penetration rate, makeup time, down time, and idle time; typically measured in ft/shift or m/shift.

Air Compressor. A device that converts power into potential energy stored in pressurized air.

Annular Space. The theoretical volume equal to annulus × length of the installation.

Annular Thickness. The radial distance between the excavated radius created by the shoe and the outside radius of the casing. The radial distance is equal to the overcut and shoe cut.

Annulus. The theoretical area between the shoe cut and the casing OD equal to [π × (shoe cut OD2 – casing OD2)/4]. The area is the combined area created by the overcut and shoe cut.

Auger Boring. *See* Horizontal Auger Boring (HAB).

Backfill Grouting. Grout injected into the void space between the carrier pipe and casing after the drive is completed.

Backstop. Also thrust block or dog plate; a reinforced area of the launching shaft wall directly behind the track or where the jacking loads will be resisted.

Band. A ring of steel welded to reinforce the casing.

Bank Cubic Yard (BCY). The calculation or measurement of 1 cubic yard3 of earth in its natural state before it is removed from the ground.

Bentonite. Expandable clay sold under various trade names that forms a slick slurry or gel when water is added; also referred to as lubricant.

Blow. Single impact of the hammer that generates energy.

Bore. An underground hole, typically horizontal, produced by tunneling or horizonal auger boring, primarily for the purpose of installing services.

Boring. (1) The dislodging and displacement of spoils by a tunnel-boring machine (TBM), rotating auger, or drill string to produce a hole or bore. (2) Obtaining geotechnical samples for evaluation and testing.

Boring Pit. *See* Pit.

Boulder. According to ASTM D653, a rock fragment, usually rounded by weathering or abrasion, with a dimension of 12 in. (305 mm) or more.

Bulked. Increased volume of earth after excavation from its natural state.

Buried Objects. Items in the ground that can affect casing advancement and line and grade control. Buried objects that prevent continued advancement along the design path including tolerances are obstructions.

Carbon Footprint. The amount of carbon dioxide generated and caused by a construction project including any traffic impacts.

Carrier Pipe. Pipe for conveyance for water, gas, sewage, or other products or services. Some carrier pipe can be installed by direct jacking (i.e., one-pass installation).

Cased Bore. A bore in which a pipe, usually a steel casing, is inserted simultaneously with the boring operation. Usually associated with horizontal auger boring or pipe ramming.

Casing. A direct-jacked or installed pipe in a two-pass system that supports a bore. The casing is not a carrier pipe. In pipe ramming, the casing is always steel.

Casing Resistance. Resistance to advancement caused by earth pressure and associated frictional drag around the pipe or casing.

Cemented Soils. Soils in which the particles are held together by a naturally occurring or human-made chemical agent that changes the soils' physical and mechanical properties.

Centerline. A real or imaginary line through the center of something, especially one following an axis of symmetry. In a trenchless application, the centerline is through the vertical center of the pipe or boring/ramming equipment.

Circumferential. The perimeter around the inner or outer surface of a circular pipe cross section.

Closed Face. Term referring to one mode of operation of a traditional tunnel boring machine. For pipe ramming, refers to a smaller diameter pipe with a closed cone-shaped end.

Coating. A protective layer applied to the outside of a pipe.

Cobble. According to ASTM D653, a rock fragment, usually rounded or semirounded, with dimensions between 3 and 12 in. (75 and 305 mm).

Cohesionless Soils (Noncohesive Soils). Earth materials containing less than 20% soil particles passing the #200 sieve. In addition, the Atterberg limits test must have a plastic index less than 7 and a liquid limit less than 25. Any free running soil, such as sand or gravel, whose strength depends on friction between particles.

Cohesive Soils. Earth materials containing 20% or more soil particles passing the #200 sieve. In addition, the Atterberg limits test must have a plastic index greater than 7 and a liquid limit greater than 25.

Collar. A fabricated collar made from steel, that is mounted on the end of the casing string. This collar is intended to transfer the ramming force from the hammer to the casing.

Collet. Segmental collar with fabricated steel segments that have a lip, which is placed inside the end of the casing, to transfer the energy from the hammer to the casing. A collet can be made up of three to four pieces for ease of handling.

Conductor Casing. Inclined steel pipe installed prior to the horizontal directional drilling process to extend below the surface into stable ground to manage the drilling fluid.

Conduit. A broad term that can include pipe, casing, tunnel, culvert, drain, duct, or channel for conveying water or other fluids or for housing cables and wires.

Cone. A fabricated steel cone with an opening(s) mounted on the end of the casing string. This cone is intended to transfer the ramming force from the hammer to the casing. It comes in different shapes and sizes. Depending on the manufacturer and region of the country, the cones are known as drive cones, ram cones, or reducing cones.

Consume. Process whereby an existing pipeline or culvert is completely engulfed or swallowed by a new or replacement installation.

Contact Grouting. Grout injected into the theoretical space between the casing and ground after the drive is completed.

Contaminated Plume. The underground trace of an environmental pollutant.

Contract Documents. Documents prepared by the owner and project engineer for bidding and for awarding a project; they can include bid forms, general conditions, special conditions, technical specifications, drawings, geotechnical data reports, and geotechnical baseline reports.

Cradle. A structure constructed from steel, concrete, or masonry that provides structural support to a hammer, casing, or carrier pipe. It typically surrounds the bottom and the sides of the object supported up to the spring line.

Cradle Machine. A horizontal auger boring machine typically supported by another machine that uses winches to advance the casing.

Crossing. Pipe installation in which the primary purpose is to provide one or more passages beneath a waterway, road, or other surface obstruction.

Crown. The highest inside surface or point of the pipe or tunnel along its total length.

Cut-and-Cover. *See* Open Cut.

Damping. A decrease in the amplitude of an oscillation as a result of energy being drained from the system to overcome frictional or other resistive forces.

Dead Man. A fixed anchor point used in advancing a saddle or cradle machine.

Delay Time. Lost time associated with the owner's issues.

Design Specification. A document that establishes specific requirements that the contractor must use, including the means and methods. A design specification creates an implied warranty that if the contractor uses the specified means and methods along with industry-accepted good practices, the constructed product will meet the specification requirements.

Desktop Study. A compilation and review of available site, geotechnical, and other information pertinent to a project. Typically performed at the outset of a project, prior to field explorations and detailed design.

Dewatering. The act of removing groundwater or lowering the groundwater elevation, using a system of wells and pumps.

Differing Site Conditions (DSC) Clause. A clause included in the contract document to address how unanticipated ground conditions are handled.

Dog Plate. *See* Backstop.

Dogs. Movable protrusions on the jacking system that engage holes or blocks in the track.

Downtime. Time lost when the hammer is unable to operate; usually associated with equipment or system failure.

Drilling Fluid or Mud. Water that may contain bentonite, polymers, soda ash, surfactant, or other materials to enhance stability and excavatability. Typical trenchless methods that use drilling fluid include microtunneling and horizontal directional drilling.

Drivability Graph. A plotted curve indicating the variation of driving resistance in force as a function of driving resistance in ramming hammer blows per foot of penetration.

Drive. Designation of the pipe installed from a launching shaft to a receiving shaft.

Drive Cone. *See* Cone.

Drive Energy. Hammer energy × efficiency.

Drive Shaft. *See* Launch Shaft.

Dry Bore. Any drilling or rod pushing system that does not use drilling fluid in the process. Usually associated with guided impact moling but also some rotary methods.

Duct. (1) In many instances, a term interchangeable with pipe; (2) In the boring industry, usually used for small plastic or steel pipes that enclose wires or cables for electrical or communication uses; (3) Conduit inside which a utility service is installed.

Earth Piercing. (1) Term commonly used in trenchless technology industry as an alternative to impact moling; (2) The use of a tool that includes a percussive hammer within a suitable casing, typically torpedo shaped. The hammer may be pneumatic or hydraulic. The term is usually associated with nonsteered devices without rigid attachment to the launch pit, relying on the resistance (friction) of the ground for forward movement. During operation, the soil is displaced, not removed. An unsupported bore may be formed in suitable ground, or a pipe is drawn in, or pushed in, behind the tool. Cables may also be drawn in.

Earth Plug. *See* Plug.

Effective Stresses. In a saturated mass of ground or rock, the total stress P minus pore water pressure Uw.

Efficiency. Amount of energy transferred to the casing as a percentage of the energy delivered by the hammer.

Embedment. A process to embed the leading edge of the casing in the ground to help eliminate wave propagation effects on the longer assembled casing string.

Energy Transfer Ratio (ETR). The ratio of energy transferred to the casing and maximum (manufacturer) rated energy for the pipe ramming hammer.

Entrance/Entry Shaft. *See* Launching Shaft.

Exit Shaft. *See* Receiving Shaft.

Face. The location where excavation is taking place. The wall of the launching shaft into which the bore is made.

Face Resistance. Resistance to penetration of the leading edge of the casing.

Factor of Safety. A ratio between the structural capacity of a system to expected or actual applied loads on the system.

Force. (1) Strength or energy as an attribute of physical action or movement. (2) Mass × acceleration.

Gravel. According to ASTM D653, rounded or semirounded particles of rock that will pass a 3 in. (75 mm) US standard sieve and will be retained on a No. 4 (4.75 mm) US standard sieve.

Ground. Earth material below the surface in which the pipeline is installed. Ground can be soil, rock, or some combination thereof.

Ground Cover. Distance between the ground surface and the crown of the casing.

Groundwater. All subsurface water as distinct from surface water.

Grouting. The process of filling voids or modifying/improving ground conditions. Grout materials may be cementitious, chemical, or other mixtures.

Guidance System. System that locates/monitors the actual position of the casing relative to the design line and grade. Systems include Dutch water level (grade only) or pilot tube.

Guide Rail. A device used to support or guide the casing within the launching shaft during a pipe ramming operation. The rails can also be used with a horizontal auger boring machine for the removal of the spoils within the casing.

Guided Boring. A general term used to describe multiple trenchless methods of installing a pipe to a desired line and grade using a guidance system. The boring process does not provide continuous support at the face of the excavation to balance groundwater and earth pressures.

Guided Pipe Ramming. A term used to describe pipe ramming with the use of a pilot tube to establish a design line and grade. Upsizing adaptors can be used in series to go from the pilot tube to the final casing diameter.

Hammer. A pneumatic or hydraulic device that imparts energy to drive the casing into the ground.

Hammer Energy. Amount of energy available for driving after internal and support system energy loss.

Hammer-Pipe Connection. A series of elements, typically steel, that are used to connect the pipe ramming hammer to the casing, and may include one or more of collets, collars, pipe cone, or a ram cone. The hammer-pipe connection may include one or both of tightening chains or a winch-tensioned cable to improve energy transfer.

Heave. Measurable upward movement of the ground or structure as the result of the casing installation.

Horizontal Auger Boring (HAB). A technique for forming a bore from a jacking or drive shaft to a receiving shaft by means of a rotating auger with cutting tools. The casings are jacked forward sequentially in a cyclic process while turning the auger. Spoils are removed back to the drive shaft by helically wound auger flights rotating in the steel casing. The equipment may have limited guidance and steering capability.

Horizontal Directional Drilling (HDD). A surface-launched trenchless technology for the installation of pipes, conduit, and cables; HDD creates a pilot bore along the design pathway and reams the pilot bore in one or more passes to a diameter suitable for the insertion of the carrier pipe. The carrier pipe is pulled into the prepared bore in the final step of the process. HDD is typically used in nongravity flow applications.

Instrumentation. Specific tools installed or used to measure and provide information on the performance of the pipe ram envelope. Instrumentation is used to measure settlement, heave, pore pressures, strains, accelerations, and ground vibrations.

Invert. Lowest inside surface or point of the pipe or tunnel along its total length.

Jacking. The actual pushing of pipe or casing in an excavated hole. This is usually done with hydraulic cylinders (jacks), but it has also been accomplished with mechanical jacks and air jacks.

Jacking Forces. The total force required to overcome the frictional resistance and face pressure components along the pipe to allow the forward movement of the boring equipment and trailing pipe string. For pipe ramming, it is only the pipe string.

Jacking Pipes. Pipe sections with smooth outside joints designed to be installed using pipe jacking techniques.

Jacking Shaft. *See* Launching Shaft.

Kinetic Energy. ½ mass × velocity2.

Launching Shaft. A vertical excavation (such as a rectangular excavation for which length is slightly more than a pipe section length and depth is slightly more than pipe invert) from which trenchless technology equipment is launched for the installation by pushing or ramming.

Launch Seal. *See* Seal.

Length. Distance between two points.

Lining. A protective layer applied to the inside of a pipe.

Lubrication. Injection of a fluid, normally bentonite and/or polymer, used to reduce the skin friction and jacking forces on the casing during installation. The fluid fills the annular space.

Makeup Time. The time spent to add a casing and make connections.

Man Entry. *See* Personnel Entry.

Mass. Force per acceleration of gravity.

Maximum Allowable Jacking Force. The largest jacking force that can be applied to the casing, allowing for an appropriate factor of safety.

Maximum Anticipated Jacking Force. The largest anticipated jacking force required to advance the casing.

Microtunneling. A trenchless method to install pipe into the ground that meets four specific features during construction. (1) Remote controlled: The microtunneling boring machine (MTBM) is operated from a control panel, normally located on the surface. The system simultaneously installs pipe as the ground is excavated and removed. Personnel entry is not required for routine operation. (2) Guided: The guidance system usually references a laser beam projected onto a target in the MTBM, capable of installing gravity sewers or other types of pipelines to the required tolerance for line and grade. (3) Pipe jacked: The process of constructing a pipeline by consecutively pushing pipes and the MTBM through the ground using a jacking system for thrust. (4) Continuously supported: Continuous pressure is provided to the face of the excavation to balance groundwater and earth pressures.

Midweld. Two or more pipe sections welded to form a longer pipe section.

Mixed Face. An interface within the pipe ram horizon between two geological units with significant contrast in engineering properties (e.g., rock overlain by soft ground or very soft, low-strength soil overlain or underlain by a very stiff, high-strength soil).

Mixed Ground. The pipe ram horizon without a defined interface that contains soil and coarse fractions of gravels, cobbles, and boulders (i.e., silt with boulders). This term is not currently defined by ASTM International or the United Soil Classification System (USCS).

Mixed Reach. A significant change in ground type, characteristics, or behavior along the pipe ram horizon between the launching and receiving shafts.

Moling. *See* Earth Piercing.

Muck. Spoils or removal of same.

No-Dig Technology. *See* Trenchless Technology.

N-Value. The number of blows per foot required to drive a standard penetration test (SPT) soil sampler into the ground during geotechnical exploration (ASTM D1586). The N-value is often used to estimate in situ consistency of clays and silts and relative density of sands.

Noncohesive soils. *See* Cohesionless Soils.

Obstruction. Any buried object that lies completely or partially within the cross section of the pipe ram envelope that impedes continued forward progress along the design path.

One-Pass. Installing a carrier pipe directly into the ground by jacking.

Open Cut (Conventional Trenching). A method by which access is gained to the required level underground for the installation, maintenance, inspection, repair, or replacement of a pipe, conduit, or cable. The excavation is then backfilled and the surface restated.

Open Face. Term referring to a mode of operation of a traditional tunnel boring machine.

Overcut. The theoretical difference of the radial measurement of the shoe cut and the casing and is equal to (shoe cut OD – casing OD)/2. Actual overcut is reduced as the shoe is worn.

Owner. A party that possesses the exclusive right to hold, use, benefit from, enjoy, convey, transfer, and otherwise dispose of the project.

Peak Vector Sum (PVS). The maximum value of the square root sum of squared longitudinal, transverse, and vertical particle velocities from a velocity time history for a given hammer blow, usually computed when monitoring construction-induced vibrations.

Penetration Rate. Instantaneous advance of the casing per time while the hammer is operating, typically measured in in./min or mm/min.

Perched Water. An accumulation of groundwater that is above the regional groundwater table in an unsaturated zone.

Performance Specification. Establishes the performance criteria the constructed product must meet. A performance specification leaves the selection of the means and methods to the contractor.

Personnel Entry. Describes any inspection, construction, renewal, or repair process, which requires an operator to enter a pipe, duct, tunnel, or bore. OSHA currently has no minimum size limit for worker-entry operations; however, a much broader concept of confined space is addressed in 29 C.F.R. 1910.146. The minimum size for which this is currently permissible in the United Kingdom is 36 in. (915 mm). Many trenchless technologies, such as the pipe ramming method, generally do not require worker entry inside the pipe.

Piercing Tools. *See* Earth Piercing.

Piezometer. A specialized monitoring well that provides discrete screens and underground seals to provide a measure of the groundwater pressure at specific intervals below ground. The groundwater pressure may be measured using open stand pipes or with vibrating wire transducers.

Pig. A device inserted into a pipeline to clean the pipe when propelled by air, water, or fluid stream pressure.

Pile Driving Analyzer (PDA). System used to monitor the real-time installation of the casing, including penetration rate, transfer energies, and installation stresses.

Pilot Tube. A small-diameter, hollow, stiff-walled pipe string (tube). Typically, a light source is mounted at the leading edge and pushed through displaceable ground. Insertion of the tube is the first step in the pilot tube method or other methods where the pilot tube is used for guidance.

Pilot Tube Method (PTM). A multistage method of accurately installing a pipe to line and grade by use of a guided pilot tube followed by upsizing to install the pipe.

Pipe Crushing. Process by which an existing corrugated metal pipe (CMP) culvert is completely engulfed or consumed by a new larger diameter casing. During the consumption process, the CMP culvert is crushed into a cloverleaf pattern for extraction.

Pipe Extraction. Process by which an existing old pipeline is replaced/extracted from the ground while at the same time a new pipeline is inserted into the hole vacated by the old pipeline.

Pipe Jacking. A system of directly installing pipes behind a shield machine by hydraulic jacking from a jacking shaft such that the pipes form a continuous string in the ground.

Pipe Lubricant. *See* Lubrication.

Pipe Pusher or Puller. A machine that pushes or pulls a rod or pipe to produce a bore by means of compaction without rotation or impact.

Pipe Ram Envelope. The vertical band within the ground where the casing is driven and includes one casing diameter above the crown and one casing diameter below the invert.

Pipe Ramming. A trenchless installation method whereby a percussive hammer is attached, via an adaptor, to an open-ended casing, which is driven through the ground. The spoils within the casing are removed after the drive is completed, or periodically during the drive, to create an open casing.

Pipe String. The succession of joined individual pipes being used to advance and support the excavation.

Pipe Swallowing. Process of consuming an existing pipeline with a new installation.

Pit. A shallow excavation used for launching or receiving the trenchless equipment. A pit typically does not have an engineered support system.

Pitch. The upward or downward angle of the casing measured from a theoretical horizontal plane passing through the longitudinal axis of the hammer.

Plowing. The act of moving earth forward during the trenchless installation process.

Plug. Intentional buildup of material in the leading section of the casing; intended to improve face stability. In flowing soil conditions, an artificial plug may be constructed. Also known as an earth or soil plug.

Porewater Pressure. Refers to the pressure of the groundwater held within the ground in gaps between particles (pores).

Port. Opening in the barrel of the casing used to introduce fluids (lubricant and/or grout) into the annular space between the casing and excavated ground.

Portal. A horizontal point of entry into the ground typically consisting of three supported cut faces.

Potholing. Small, carefully dug, nonmechanized excavation to locate a utility or other subsurface feature.

Power. Work per change in time.

Product Pipe. *See* Carrier Pipe.

Project. An individual or collaborative enterprise that is carefully planned and designed to achieve a particular aim.

Pullback. A step in the installation process in which the carrier pipe is pulled into the created hole or casing.

Quake. The magnitude of relative soil-casing movement when the assumed soil resistance force-relative displacement relationship transitions from an elastic response to a perfectly plastic response. This magnitude of displacement must be specified in wave equation analyses of drivability.

Ramming. A percussive hammer is attached to an open-end casing, which is driven through the ground. *See* Pipe Ramming.

Rebound. Situation in which the casing bounces back after hitting the ground during initial entry or a buried object.

Receiving Shaft. Excavation into which the trenchless technology equipment and pipe string are driven and recovered.

Refusal. Point at which the casing no longer advances.

Rescue Shaft. An unplanned additional shaft to remove obstacles/obstructions.

Rock. According to ASTM D653, any naturally formed aggregate of mineral matter occurring in large masses or fragments [International Society of Rock Mechanics (ISRM)].

Roll. The angle of rotation about the theoretical longitudinal centerline of the pipe string measured from the 12 o'clock position.

Seal. A mechanical seal, usually consisting of a rubber flange (can be reinforced with slide plates) that is mounted to the interior wall of shaft. The flange seal is distended by the casing as it passes through. The seal is intended to prevent groundwater, soils, and lubrication inflow into the shaft during installation.

Settlement. Measurable downward movement of the ground, overlying utility, or other structure as the result of the excavation process or dewatering.

Settlement Point or Marker. An instrument installed at ground surface or driven into the ground to monitor vertical ground deformations before, during, and after construction.

Shaft. (1) A passage made from the surface of the ground to a point underground, the longer axis of which makes an angle greater than 20 degrees to the horizontal; (2) a pit in which there are employees, and it is foreseeable that they may enter (or do enter) the horizontal excavation; or (3) a pit that has typical underground construction hazards and is connected to a horizontal excavation.

Shoe. A ring of steel welded at the front of the lead section of the casing—either inside, outside, or both—to cut through the ground, to create relief (overcut), and to reinforce and strengthen the casing (used also in horizontal auger boring). Tooling can be added to the shoe depending on the ground conditions.

Shoe Cut. The outermost projection of the shoe measured as a diameter, minus the outside diameter of the casing.

Skin Friction. *See* Casing Resistance.

Slickbore. Process whereby a sacrificial casing is replaced/extracted from the ground while at the same time a new carrier pipe is inserted into the hole vacated by the casing.

Social Costs. Costs incurred by society as a result of underground pipeline construction and renewal. These include, but are not limited to, traffic disruptions, environmental damage, safety hazards, inconvenience to the general public, and business losses resulting from road closures.

Spoil. Earth, rock, and other materials excavated during the installation process.

Stiffness of Pipe. (Elastic modulus × moment of inertia)/(0.149 × nominal radius3).

Strength. The capacity of an object or substance to withstand force or pressure.

Sump Pump. A device placed at the base of the shaft or excavation to collect and remove fluids incidental to the construction process and shaft leakage and to prevent the casing from flooding.

Third Party. An entity (person, organization, agency, business, and so forth) that may be indirectly involved but is not a principal party to a project, arrangement, or contract.

Thrust. Force applied to a pipeline or drill string to propel it through the ground.

Thrust Block. *See* Backstop.

Trenching. *See* Open Cut.

Trenchless Technology. A family of construction techniques for installing or rehabilitating underground infrastructure with minimal disruption to surface traffic, businesses, and residents. Also includes technologies for inspection, leak location and detection, with minimal disruption and excavation from the ground surface.

Tunnel. An underground passage dug through the earth; built through a hill or under roads, railroads, buildings, or rivers.

Two-Pass. Installing a carrier pipe in a casing.

Uncased Bore. A bore without a lining or pipe inserted (i.e., a self-supporting hole, either temporary or permanent). Typically not recommended, except under special circumstances.

Underground Utility. Active or inactive services or utilities below ground level.

Upsizing. Process of selecting a larger diameter than originally needed. Also, any method that increases the cross-sectional area of an existing pipeline, replacing it with a larger cross-sectional area pipe.

Upsizing Adaptor. A fabricated tool to transition from the pilot tube or other immediate diameters to the final casing diameter to help transfer shear from the smaller diameter to the larger diameter.

Utility Monitoring Point (UMP). An instrument designed to measure movement of an individual utility when exposed to construction activities.

Utility Tunneling. A construction method to excavate an opening beneath the ground without continuous disturbance of the ground surface. The excavation is of sufficient diameter to allow personnel access, excavation, transport of spoils, and erection of a ground support system.

Velocity. Change in distance per change in time.

Void Space. Area between the interior of the casing and the exterior of the carrier pipe × the length of the installation; *see* Annular Space.

Voids. (1) Holes external to the pipe in the surrounding soil or material. (2) A term generally applied to paints to describe holidays, holes, and skips in the film. (3) Shrinkage in castings or welds.

Washover Casing. Similar to a conductor casing, installed before or after the horizontal directional drilling process to manage the drilling fluid or provide ground support to the drill stem or stuck pipe.

Water Jetting. Internal cleansing mechanism in which high-pressure water is sprayed from nozzles to help remove cohesive soils.

WEAP. Computer program to analyze the stresses during installation adopted from pile driving and applied to pipe ramming. WEAP stands for Wave Equation Analysis of Piles.

Work. (1) Specific activities defined in a contract document. (2) Force in the direction of movement × displacement.

Worker Entry. *See* Personnel Entry.

Yaw. The angle of the pipe string to the left or right of a theoretical vertical plane passing through the longitudinal axis of the shoe.

Zone of Influence. Volume of ground that could possibly be impacted by settlement or heave from either the tunnel mining, shaft excavation, pipe ramming, pile driving, or other construction activities.

INDEX

Page numbers followed by *e*, *f*, and *t* indicate equations, figures, and tables.

abandoned utilities, site investigations and, 100
accuracy, design considerations and, 122
adjacent structures: direct costs and, 79; planning and, 74–75
air compressor, equipment selection and setup, 19–20
alignment and distance considerations, planning and, 74
American Railway Engineering and Maintenance-of-Way Association (AREMA), 24, 44, 113, 123
annular space backfill, 38–40
applications for pipe ramming: combining pipe ramming and horizontal directional drilling, 49–53, 51*f*, 52*f*, 53*f*, 54*f*; culverts, 44–47, 45*f*, 46*f*; pipe crushing, 9, 47, 48*f*; pipe extraction, 49, 50*f*; road and railroad crossings, 5*t*, 43–44; slick boring, 48–49, 49*f*
as-built drawings and documentation, 150
ASCE 38-02, 67–68, 100

backlogs, direct costs and, 77
barricades, construction and, 146
bedrock, site investigation and, 91
bentonite lubricants, 13, 33
bidding, 131–132
bore (product pipe) salvage, 50, 51*f*
boulder volume ratio (BVR), 96
buried objects: design considerations and, 122; direct costs and, 77; planning and site conditions, 68; site investigation and potential, 96. *See also* utilities
business disruptions: planning and, 71; social effect costs, 80–81
bypass pumping, planning and, 72

caliche, site investigation and, 97
carbon footprint, 64–65, 81
carrier pipe installation, equipment selection and setup, 37–42, 39*f*
case histories, 153–154, 153*f*, 154–164*t*
case pieces, joining of, 33–36, 35*f*, 37*f*
casing penetration and advancement, equipment selection and, 29–32, 31*f*
casing size, design exploration

and, 120
casings: design considerations, 25*t*, 113, 120, 122–123, 126–127, 127*e*, 128*t*; direct costs, 76, 78; equipment selection and setup, 24, 25*t*, 26, 31*f*; initial planning criteria, 58–59; in preconstruction submittals, 133; steel casing, leading edge treatments, and welding, 144
C-channel, equipment selection and setup, 14, 15*f*
cellular grout, 38, 41
cemented soil, site investigation and, 97
closed-face process (piercing, impact moling, compaction), 11–12, 13
cobble volume ratio (CVR), 96
collets, 26–29, 27*f*, 28*f*, 29*f*
competing projects, direct costs and, 77
conductor/washover casing, 52–53, 54*f*
cone penetration tests (CPTs), 90
construction: bidding, 131–132; as-built drawings and documentation, 150; differing site conditions (DSC), 132, 145–146; inspection and monitoring, 149; jobsite layout and equipment setup, 140–141, 142*f*; launching and receiving shafts, 142–144, 142*f*; measurement and payment, 137–140; quality control, 146–147; ramming force and lubricants, 147–148, 148*f*; reports and records, 150, 151*f*; safety issues, 145; spoils removal and disposal, 148–149; steel casing, leading edge treatments, welding, 144; submittals, 132–137; survey, 141–142; traffic control, fencing, and barricading, 146
construction moratoriums, planning and, 72
construction submittals, 136–137
consumption (pipe swallowing), 9, 45–46, 45*f*, 47, 48*f*
contaminated ground: planning and, 60; site conditions and, 68–69; site investigations and, 97, 102
contaminated water, site investigation and, 97
contingency costs: design and, 110; planning and, 84–85
contingency plans, in preconstruction submittals, 135–136
contractor concerns, hazard identification and design, 111–112
correction ability, design considerations and, 122
corrugated metal pipe (CMP) culvert, 47, 48*f*
costs considerations, 75; design and, 109–110; direct costs, 76–80; environmental costs, 82–83; indirect social costs, 80–82; risk analysis, 83–84, 84*f*
culverts, 44–47, 45*f*, 46*f*
cutting ring, in open-face process, 12–13
cutting shoes, 13, 20–24, 21*f*, 22*f*, 23*f*

daily report, 151*f*
deformation monitoring points (DMPs), 149
depth: direct costs, 78; equipment selection and setup, 16–17; initial planning criteria, 59–60
design, 107–108; design considerations, 121–128, 125*f*, 127*e*, 128*t*; documentation, 128–129; exploration, 114–121; planning and permitting, 108–114

desktop study: geotechnical project aspects, 73; site investigation and, 88–89
detailed site investigations, 87; environmental conditions, 101–102; flood zones, 103–104; geotechnical assessment, 87–98, 93–95*t*, 99*t*; seismic considerations, 104; traffic flow and access for vehicles and pedestrians, 100–101; utility surveys, 100
dewatering, 32, 77, 96, 98, 116, 131, 135, 140, 144
differing site conditions (DSC), 132, 145–146
dimension ratio (DR), 123
disposal sites, planning and, 71
drill stem recovery, 51, 52*f*
drive cones, 26–27, 29
dust control: planning and, 70; social effect costs, 81

easements and right-of-way, planning and site conditions, 66–67
Electric Power Research Institute (EPRI), 40
engineering concerns, hazard identification and design, 111
environmental baseline report (EBR), 69
environmental conditions: costs and, 82–83; restrictions and, 70; site investigation and, 97, 101–102
equipment selection and setup: air compressor, 19–20; carrier pipe installation, 37–42, 39*f*; casing selection and wall thickness, 24, 25*t*, 26; construction and, 140–141, 142*f*; cutting shoes, 13, 20–24, 21*f*, 22*f*, 23*f*; depth, 16–17; direct costs, 78; efficiency of casing penetration and advancement, 29–32, 31*f*; guide rails and setting grade, 13–14, 15*f*, 16*f*, 118*f*; hammer attachment, 16*f*, 17*f*, 18*f*, 19*f*, 26–29, 27*f*, 28*f*, 29*f*; hammer selection, 17, 17*f*, 18*f*, 19, 19*f*; joining casing pieces, 33–36, 35*f*, 37*f*; minimum space requirements, 14–16, 17*f*, 18*f*, 19*f*; spoils removal, 36
exploration, design and: casing size, 120; geotechnical considerations, 116–117; ground settlement or heave, 117–119; groundwater, 117, 118*f*; line and grade deviations, 119–120, 119*f*; planning for, 114; ram length limitations, 16*f*, 17*f*, 120–121; site layout, 114–116
extraction cradle, 50, 51*f*

fabricated sleeve, 50
fencing, construction and, 146
flood zones: contingencies for contractor, 104; site investigations, 103–104
flow capacity, 20, 45–47, 58
forward propagation, 125
foundation stability, 14
free-flowing sand, 32
freehand field cutting, of casing, 36

geotechnical assessment, site investigations, 87–88; conditions and applicability of ramming, 98, 99*t*; construction and differing conditions, 132, 145–146; geotechnical reports, 97–98; phases of, 88–92; subsurface materials characterization, 92–93, 93–95*t*, 96–97
geotechnical baseline report (GBR), 98
geotechnical considerations, 64;

design, 112–113, 116–117; direct costs, 76; planning, 73; site investigation, 97–98
geotechnical data report (GDR), 97
geotechnical interpretive report (GIR), 97–98
grade: direct costs and, 77–78; initial planning criteria, 58–59
gravity sewers, 58–59
ground conditions, site investigation and, 92–93, 93–95*t*, 96–97
ground response, design considerations and, 123–126, 125*f*
ground settlement or heave, design exploration and, 117–119
ground shear failure, 8
ground stabilization, in preconstruction submittals, 136
groundwater: design and, 113, 117, 118*f*; direct costs, 77; equipment selection and setup, 16; planning, 60, 69, 72; site investigation, 93–95*t*, 96; spoils and, 32
grout backfill, 39–40
grout mixing, 41–42
guide rails and setting grade, equipment selection and setup, 13–14, 15*f*, 16*f*, 27, 118*f*

hammer attachment, equipment selection and setup, 16*f*, 17, 17*f*, 18*f*, 19, 19*f*, 26–29, 27*f*, 28*f*, 29*f*
hammer energy, 7, 20, 24, 27–28, 29, 116, 122
hammer ground dynamics, 8
Hammerhead Trenchless, 10
haul routes, planning and, 71
hazard identification, design and, 108–109, 110–113
heave: design considerations, 123–126, 125*f*; design exploration, 117–119
high risk hazards, design and, 113
high-density polyethylene (HDPE) pipe, 8
horizontal auger boring system (HAB), 148
horizontal directional drilling (HDD), 49–53, 51*f*, 52*f*, 53*f*, 54*f*
hydraulic hammers, 8
hydrolock, 52

I-beam, equipment selection and setup, 14, 15*f*
IHC Hydrohammer, 10
in situ vane testing, 91
inclinometers, 149
initial planning criteria, 57–58; casing diameter and grade, 58–59; depth, 59–60; portals, shafts, and working space, 5*t*, 61–64, 62*f*; regulations, 65; routing, 60–61; sustainability, 64–65
inner and outer shoes, 144
inspection and monitoring, of construction, 149

jobsite layout and equipment setup, 140–141, 142*f*
joining case pieces, equipment selection and setup, 33–36, 35*f*, 37*f*
joint configuration, design considerations, 126

labor conditions, direct costs, 77
land use, planning and site conditions, 65–66
lateral propagation, 125
launching and receiving shafts: construction, 138–139, 142–144, 142*f*; design, 110, 115; planning, 61–64; site investigation, 89, 96, 102, 104
layout of project, 140–141, 142*f*; alignment and distance considerations, 74; design exploration

and, 114–116; effects on adjacent structures, 74–75; for efficient ramming, 73–74; in preconstruction submittals, 133–134
leading edge treatments, 144
light limitations, planning and, 70–71
line and grade, design and, 119–120, 119*f*, 121–122
location, direct costs and, 78
lubricants and lubrication, 13, 22–23, 32–33, 34*f*, 134, 147–148, 148*f*
lump sum payments, 109–110, 137–138

Manual on Uniform Traffic Control Devices (FHWA), 146
measurement and payment, construction and, 137–140
minimum ground cover, planning and, 59–60
minimum space requirements, equipment selection and setup, 14–16, 17*f*, 18*f*, 19*f*
mobilization, direct costs and, 78

noise concerns, 9, 32, 70, 81, 97, 141, 146

Occupational and Health Safety Administration (OSHA) regulations, 63
odor concerns, 71, 82
one-call ticket, 67
open trench excavation, 1
open-face process, 11–13
owner concerns, hazard identification and design, 111

parallel construction, culverts and, 46–47, 46*f*
pavement restoration, social effect costs, 82
payment arrangement, 137–140
pedestrian disruption concerns, 80, 100
penetration rate, 8, 30–32, 116, 150
Permalok, 35, 144
personnel qualifications, in preconstruction submittals, 133
photoionization detector (PID), 97
piezometers, 149
pilot tube guidance, 42–43, 43*f*
pipe crushing, 9, 47, 48*f*
pipe extraction, 9, 49, 50*f*
pipe ramming process: applicability of ramming, 98, 99*t*; appropriate installations, 8–9; background of, 9–11, 11*f*; closed-face process (piercing, impact moling, compaction), 11–12, 13; components of, 29; introduction to, 7–9; open-face process, 11–13; pilot tube guidance and, 42–43, 43*f*; typical applications, 43–53. *See also* equipment selection and setup
pipe swallowing (consumption), 9, 45–46, 45*f*, 47, 48*f*
Pipeline Route Selection for Rural and Cross-Country Pipelines (ASCE), 60
planning: direct costs, 76–80; environmental costs, 82–83; geotechnical considerations, 73; indirect social costs, 80–82; initial criteria, 57–65; project layout, 73–75; risk analysis, 83–84, 84*f*; site conditions, 65–72
planning and permitting, design and: cost and schedule, 109–110; hazard identification, 108–109, 110–113; high risk hazards, 113; permitting and design for special crossings, 113–114; process, 108–109
plowing, 32

pneumatic hammers, 8
polymer lubricants, 13, 32–33
portals, shafts, and working space, initial planning criteria, 5*t*, 61–64, 62*f*
postconstruction submittals, 137
pre-bid meetings, 132
precision, design considerations and, 121–122
preconstruction submittals, 133–136
premium projects costs, design and, 110
project complexity, direct costs and, 77
proof of concept, 108
public events, planning and, 72
public safety. *See* safety issues
pullback assist, 52, 53*f*

qualified bidders, direct costs and, 77
quality control, construction and, 146–147
Quality Level records searches, 67

ram length limitations, design exploration and, 16*f*, 17*f*, 120–121
ramming force, 147–148, 148*f*
receiving shafts. *See* launching and receiving shafts
recoil, 30, 31*f*
redline drawings, 150
refusal, 31–32, 36
regulations: design and, 110–111; initial planning criteria, 65
rehabilitation, culverts and, 44–45
replacement, culverts and, 45–46, 47
reports and records, 150, 151*f*
reputation of owners, direct costs and, 79
rescue methods, 139
residential structures: limited work hours, 70; social effect costs, 81
restricted access, planning and, 70
risk analysis, planning and, 83–84, 84*f*
road and railroad crossings, 5*t*, 43–44
rock conditions, site investigation and, 93, 95–96
roller cutters, 23
routing, initial planning criteria, 60–61

safety issues: construction and, 145; in preconstruction submittals, 136; social effect costs, 82
schedule, design and, 109–110
seismic considerations, site investigations and, 104
service interruptions, planning and, 71
settlement, of existing facilities, 71–72
settlement, of ground: design considerations, 123–126, 125*f*; design exploration, 117–119
setup. *See* equipment selection and setup
shafts: design in preconstruction submittals, 135; initial planning criteria, 5*t*, 61–64, 62*f*. *See also* launching and receiving shafts
sheet pile, equipment selection and setup, 14, 15*f*
site conditions, planning and, 65; buried objects, 68; construction conditions differing from, 132, 145–146; contaminated ground, 68–69; easements and right-of-way, 66–67; land use, 65–66; special considerations, 69–72; underground utilities and other facilities, 67–68
site investigations. *See* detailed site investigations

site restoration, 139
site visits, pre-bid, 132
slick bore casing, 11, 11*f*, 26, 48–49
slick boring, 48–49, 49*f*
social effect costs, 80–82, 110–111
soil port, 29
sonde, 23–24
special crossings, design and, 113–114
split spoon samples, site investigation and, 90–91
spoils removal and disposal, 36, 148–149; direct costs and, 79; planning and, 72; in preconstruction submittals, 133, 134
Standard Guidelines for the Collection and Depiction of Existing Subsurface Utility Data (ASCE), 61
standard penetration test (SPT), 90
structural monitoring points (SMPs), 149
submittals, of contractors, 132–137
subsurface easement, 66–67
subsurface investigation methods, 89–92
subsurface materials, characterization during site investigation, 92–93, 93–95*t*, 96–97
surveys, 136, 141–142
sustainability, initial planning criteria, 64–65
system efficiency, defined, 8

tax revenue, social effect costs, 81
test boring, site investigation and, 89–90
time, direct costs and: makeup time, 78–79; time of year, 79–80
traffic control: during construction and, 146; planning and, 71; site investigations and, 100–101; social effect costs, 80
trenchless excavation: basic design processes, 3; capabilities of methods, 4, 5*t*; guidelines for, 3–4; types and benefits of, 2
Tri-lock, 35, 144
TT Technologies, 10
Tunnelman's ground classification system, 93, 93–95*t*

utilities: abandoned, 100; direct costs and, 79; jobsite layout and equipment setup, 140; planning and, 61, 72; underground utilities and other facilities, 67–68
utility monitoring points (UMPs), 149
utility surveys, site investigations and, 100

vehicle access, site investigations and, 100
vibration: design considerations, 123–126, 125*f*; as work issue, 28–29, 32, 60, 63, 97, 100, 116–118
visual impact, planning and, 71

wall thickness: direct costs, 76; equipment selection and setup, 24, 25*t*, 26
washover casing, 52–53, 54*f*
weather, direct costs and, 79–80
welding, 33–36, 34*f*, 115, 126, 142, 144–145
work, defined, 8
work restrictions, 70
work plan, in preconstruction submittals, 133
working space, initial planning criteria, 5*t*, 61–64, 62*f*

zone of influence, 12, 75, 122